Stochastic and Global Optimization

Nonconvex Optimization and Its Applications

Volume 59

Managing Editor:
Panos Pardalos

Advisory Board:

J.R. Birge
Northwestern University, U.S.A.

Ding-Zhu Du
University of Minnesota, U.S.A.

C. A. Floudas
Princeton University, U.S.A.

J. Mockus
Lithuanian Academy of Sciences, Lithuania

H. D. Sherali
Virginia Polytechnic Institute and State University, U.S.A.

G. Stavroulakis
Technical University Braunschweig, Germany

The titles published in this series are listed at the end of this volume.

Stochastic and Global Optimization

Edited by

Gintautas Dzemyda

Vydūnas Šaltenis

Antanas Žilinskas

Institute of Mathematics and Informatics,
Vilnius, Lithuania

KLUWER ACADEMIC PUBLISHERS

DORDRECHT / BOSTON / LONDON

A C.I.P. Catalogue record for this book is available from the Library of Congress.

ISBN 978-1-4419-5209-7 e-ISBN 978-0-306-47648-8

Published by Kluwer Academic Publishers,
P.O. Box 17, 3300 AA Dordrecht, The Netherlands.

Sold and distributed in North, Central and South America
by Kluwer Academic Publishers,
101 Philip Drive, Norwell, MA 02061, U.S.A.

In all other countries, sold and distributed
by Kluwer Academic Publishers,
P.O. Box 322, 3300 AH Dordrecht, The Netherlands.

Printed on acid-free paper

All Rights Reserved
© 2002 Kluwer Academic Publishers
Softcover reprint of the hardcover 1st edition 2002
No part of the material protected by this copyright notice may be reproduced or
utilized in any form or by any means, electronic or mechanical,
including photocopying, recording or by any information storage and
retrieval system, without written permission from the copyright owner

TABLE OF CONTENTS

THE JUBILEE OF PROF. DR. HABIL. JONAS MOCKUS

On June 18, 2001, the scientific society commemorates the 70th anniversary of Prof. Dr. Habil. J. Mockus, a world renowned scientist, a real member of the Lithuanian academy of Sciences.

The book represents some of the results in the field of stochastic optimization, one of initiators of which was Prof. J. Mockus.

The range of J. Mockus' scientific research embraces global and discrete optimization theory, methods, algorithms, software and its applications in design, economics, and statistics. He is the author of the global optimization theory based on the Bayesian approach. In this field Jonas Mockus gained a recognized authority in the scientific world. On the basis of this theory Jonas Mockus has constructed a number of algorithms for global and discrete optimization. The methods of discrete optimization developed by J. Mockus rely on an efficient use of heuristic and approximate algorithms. The key target of work was to solve a fundamental problem that is of importance to technical sciences: how to relate heuristics found by experts with the mathematical Bayesian solution theory. A potential domain of application of the work results in theory is very wide which is withessed by a list of different problems, investigated within the framework of one theory, and each of which features important problems in practice. On the whole, solution of this very important problem required a deep insight into manysided mathematical issues, however they were only a means for increasing the efficiency of the usage of optimization methods in technical and economic systems.

Traditional mathematical methods were created with a view to ensure the accuracy. Using the guaranteed precision methods in real engineering problems, the calculation time grows exponentially, therefore to optimize large and complex systems the heuristic methods are applied. The rules of solution based on expert experience and intuition are called heuristics. Heuristic methods consume less time, however, they are not substantiated by theory, so the efficiency of their application depends on the intuition of experts.

J. Mockus has shown in his works how to use the Bayesian statistical solution theory in order to optimize the parameters of heuristic methods by randomizing and joining different heuristics. This theoretical result facilitates the improvement of heuristic methods by ensuring the convergence and essentially diminishing their average error. This equally applies both to newly created heuristics and traditional widespread heuristic optimization methods, e.g., genetic and simulated annealing algorithms.

To verify the efficiency of theoretical results, J. Mockus has studied several different mathematical models that reflect important design and control problems. In chronological order, a first application of global optimization (not only in Lithuania, but also on the international scale) was the optimization of magnetic circuit parameters of a domestic electric meter. Jointly with the Vilnius Electric Meter Plant J. Mockus has constructed its model SO–I445 that ensured the required precision even using low quality magnetic substances. A second application was the optimization of the USSR North-West Electric Power Networks to order of "Energosetprojekt", when more reliable and more economical high-voltage networks of this system and also of Lithuania have been designed. The first monograph by J. Mockus [1] presents a detailed description of this and other examples.

Under the supervision of J. Mockus, a lot of different problems of practice that reflect essential optimization problems in various technical and economical systems have been considered. The latest application of significance is working out optimal schedules for serial production in small series. The work was done in conjunction with the chemical engineering department of Purdue University and Searl Research Laboratories of the Biotechnological Company at Monsanto, USA. All this including new results in theory are described in [2]. In addition, this monograph features a direct usage of the heuristic method by means of dynamic visualization that is fit for optimizing the search for images at the statistics department of Carnegie Mellon university by analyzing 30 000 pictures of Jupiter taken by the space station "Voyager". In his monograph [3] J. Mockus describes mathematical aspects of this theory, software of global optimization as well as a number of optimization applications in practice, starting from vibroengines and finishing with nonstationary queueing systems.

Program realizations of the optimization methods developed are included into many program packages. But for the Optimization department only, headed by him, many optimization packages have been developed following various scientific programs, dependent on the computer basis and the specificity of the problems solved. All of them bear a common feature that their user can find a wide range of optimization programs: not only for global optimization, but also for local search and structural analysis of problems.

In the latest J. Mockus' monograph [4] examples of global optimization meant for the studies and research work in the medium of Internet are described.

In 1968 Jonas Mockus was awarded the State Prize of the Lithuanian SSR for the monograph [1] in Engineering. In 1970 he was elected a corresponding member of the Lithuanian Academy of sciences, in 1974 he became a real member of the same academy, and in 1998 he was given a Lithuanian Science

Prize for the monographs [2,3]. Jonas Mockus is editor-in-chief of the journal *Informatica* and a member of the editorial board of the *Journal of Global Optimization*. He is a member of the "American Mathematical Society" as well as the IFIP working group W.G.7.7 (Stochastic Optimization), and a member of Senate of the Institute of Informatics and Mathematics.

J. Mockus has prepared a generation of Lithuanian optimization specialists, most of whom are successfully proceeding in this field. Under his supervision, 18 PhD theses and 3 doctoral theses for habilitation have been maintained. J. Mockus is the author of over 100 publications including 4 monographs, as well as 87 scientific reports, 62 of which delivered at international conferences, and also he was an invited speaker and lecturer at 36 international conferences.

The subject area of J. Mockus' lectures at universities ranges within the framework of optimization methods, operations research, game theory, queueening theory, theory of statistical solutions, experiment design and reliability theory. Since 1959 Jonas Mockus has been a lecturer at the Kaunas Polytechnical Institute (presently, Kaunas University of Technology), and since 1968 – a professor of this university. From 1993 he is a professor of Vytautas Magnus University, and from 1995 – of Vilnius Gediminas Technical University. Coordination of teaching with scientific research work yields good results in preparing future specialists in informatics.

Biographic facts. J. Mockus was born June 18, 1931, in Pakruojis township in the Rokiškis district, Lithuania.

In 1947 he graduated from secondary school at Kaunas. From 1947 to 1952, the studies at the Kaunas Polytechnical Institute, Faculty of Electrical Engineering, and a speciality of elelctrical engineering acquired. From 1952 to 1956 doctoral courses at Moscow, and his successful graduation from the Institute of Energetics of the USSR Academy of Sciences.

In 1956 he defended the thesis in technical sciences maintained at Moscow, Institute of Energetics, in 1966 – a doctoral thesis defended at the Institute of Automation and Computer Engineering in Riga, in 1968, a professor's title was conferred on him, and in 1992, a doctor habilius title in informatics engineering in the field of technological sciences was conferred.

In 1957, Jonas Mockus begins his career as a senior researcher of the Energetics Institute of the Lithuanian Academy of Sciences in Kaunas, and since 1959 he has been head of the Optimization Department. Since 1970 he has been working in Vilnius as head of the Optimization Department at the Institute of Physics and Mathematics (presently, Institute of Mathematics and Informatics).

J. Mockus is married. His wife Danguole Mockiene is engaged in scientific work too. They brought up two sons – Linas and Audris – who also have chosen a scientist's career (following in their father's foot steps) and presently are successfully working at scientific institutions of the USA, keeping close scientific contacts with the research work performed by J. Mockus. 7 grandchildren are growing safe and sound. All his life J. Mockus constantly went in for one or other kind of sports: skating at his early youth, swimming and yachting at older age, a longlasting slalom in the Caucasus mountains, and skiing and hiking in Lithuanian plains so far.

List of Monographs by J. Mockus

[1] *Multimodal Problems in Engineering Design*, Nauka, Moscow, 1967.

[2] *Bayesian Heuristic Approach to Global and Discrete Optimization*, Kluwer Acad. Publ., Dordrecht, 1997, with W. Eddy, G. Reklaitis, A. Mockus and L. Mockus.

[3] *Bayesian Approach to Global Optimization*, Kluwer Acad. Publ., Dordrecht, 1989.

[4] *A Set of Examples of Global and Discrete Optimization: Application of Bayesian Heuristic Approach*, Kluwer Acad. Publ., Dordrecht, 2000.

GINTAUTAS DZEMYDA
VYDŪNAS ŠALTENIS
ANTANAS ŽILINSKAS

Prof. Dr. Habil. J. Mockus

Chapter 1

TOPOGRAPHICAL DIFFERENTIAL EVOLUTION USING PRE-CALCULATED DIFFERENTIALS

M. M. Ali
Centre for Control Theory and Optimization
Department of Computational and Applied Mathematics
Witwatersrand University, Wits-2050, Johannesburg, South Africa
mali@cam.wits.ac.za

A. Törn
Department of Computer Science
Åbo Akademi University
SF-20520, Turku, Finland
atorn@abo.fi

Abstract We present an algorithm for finding the global minimum of multimodal functions. The proposed algorithm is based on differential evolution (DE). Its distinguishing features are that it implements pre-calculated differentials and that it suitably utilizes topographical information on the objective function in deciding local search. These features are implemented in a periodic fashion. The algorithm has been tested on easy, moderately difficult test problems as well as on the difficult Lennard–Jones (LJ) potential function. Computational results using problems of dimensions upto 24 are reported. A robust computational behavior of the algorithm is shown.

Keywords: Global optimization, differential evolution, pre-calculated differential, continuous variable, topographs, graph minima

1. Introduction

The inherent difficulty of global optimization problems lies in finding the very best minimum from a multitude of local minima. We consider the problem of finding the global minimum of the unconstrained optimization problem

G. Dzemyda et al. (eds.), Stochastic and Global Optimization, 1–17.
© 2002 *Kluwer Academic Publishers.*

$$\min f(x) \quad \text{such that} \quad x \in \Omega \subset R^n. \tag{1}$$

A global minimization algorithm aims at finding the global minimizer x^* of $f(x)$ such that

$$f^* = f(x^*) \leqslant f(x), \quad \forall x \in \Omega. \tag{2}$$

Such an optimization problem involving a large number of continuous variables arises in many practical field of applications, and there is a need of designing a reliable algorithm to solving this problem.

A number of deterministic and stochastic algorithms have been proposed [1–5] for solving (1). This paper is concerned with the stochastic methods. As opposed to deterministic methods, stochastic methods use a little or no properties of the function being optimized. Moreover, the stochastic methods are easy to implement and thus preferred by many. Among the stochastic methods that lend to easy implementation are the genetic algorithm [6], controlled random search [7] and differential evolution [8]. Unfortunately these methods have theoretical limitations and they work only for low dimensional problems. The stochastic method such as simulated annealing [9,10] has convergence properties but it requires global information (analogous to the Lipschitz constant for a deterministic algorithm) in the form of an appropriate cooling schedule which may be very hard to obtain. Moreover, in a recent numerical experiment with some recent stochastic methods using practical problems it was shown that simulated annealing is the worst performer on problems with dimension as small as six [11]. Other stochastic methods are efficient multistart [12–14] and among them topographical multilevel single linkage [14] was found to be superior again on low dimensional practical problems [11]. Thus there is a desire to design algorithms which do not need global information and can be applied to 'moderately higher' to higher-dimensional problems. The purpose of this paper is to design such an algorithm which is robust and reliable in finding the global minimum for funtions of lower to higher dimensions. In designing the new algorithm, we restrict ourselves to two kinds of stochastic algorithms, namely the differential evolution (DE) algorithm of Storn and Price [8] and the topographical algorithm (TA) of Törn and Viitanen [13]. The proposed algorithm uses the complementary strengths of these two algorithms. In Section 2 we briefly describe the DE algorithm. Section 3 presents DE using pre-calculated differentials, a modified version of DE. In Section 4 the new algorithm is presented. Problems considered for numerical studies are discussed in Section 5. The results are discussed and also summarized in Section 5, and Section 6 contain the conclusions.

2. Differential Evolution (DE)

DE is a population based method. It guides an initial population set $S = \{x_1, x_2, \ldots, x_N\}$, chosen randomly from the search region Ω, to the vicinity of the global minimum through repeated cycles of mutation, crossover and acceptance. In each cycle constituting a generation, N attempts are made to replace members of S with better members for the next generation. The ith ($i = 1, 2, \ldots, N$) attempt is made to replace x_i, the target point, in S. The candidate point y_i for x_i is the result of mutation and crossover.

Mutation. The objective of the mutation operation is to obtain an intermediate point $\hat{x}_i$. It is obtained using

$$\hat{x}_i = x_{r1} + F \times (x_{r2} - x_{r3}), \tag{3}$$

where x_{r1}, x_{r2} and x_{r3} are distinct points chosen at random in S. None of these points should coincide with the current target point x_i. F is the scaling factor of the differential vector $(x_{r2} - x_{r3})$. In a recent study [15] it was found that a good value of the scaling factor, $F \leqslant 1$, is given by

$$F = \begin{cases} \max\left(l_{\min}, 1 - \left|\frac{f_{\max}}{f_{\min}}\right|\right) & \text{if } \left|\frac{f_{\max}}{f_{\min}}\right| < 1, \\ \max\left(l_{\min}, 1 - \left|\frac{f_{\min}}{f_{\max}}\right|\right) & \text{otherwise,} \end{cases} \tag{4}$$

where $l_{\min} \in [0.3, 0.4]$ and $f_{\max}$ and $f_{\min}$ respectively are the high and low function values within S [15]. If a component $\hat{x}_i^j$ falls outside Ω then it is found randomly in-between the jth lower and upper limits.

Crossover. The trial point y_i is found from the target x_i and $\hat{x}_i$ using the following crossover rule:

$$y_i^j = \begin{cases} \hat{x}_i^j & \text{if } R^j \leqslant CR \quad \text{or} \quad j = I_i, \\ x_i^j & \text{if } R^j > CR \quad \text{and} \quad j \neq I_i, \end{cases} \tag{5}$$

where I_i is a randomly chosen integer in the set I, i.e., $I_i \in I = \{1, 2, \ldots, n\}$; the superscript j represents the jth component of respective vectors; $R^j \in (0, 1)$, drawn randomly for each j. The entity CR is a constant, empirically found to be 0.5 [15]. The acceptance mechanism (whether or not y_i replaces x_i) follows the crossover.

Acceptance. In the acceptance phase a one to one comparison is made in that the function value at each trial point, $f(y_i)$, is compared to $f(x_i)$, the value at the target point. If $f(y_i) < f(x_i)$ then y_i replaces x_i in S, otherwise, S retains the original x_i.

This process of targeting x_i and obtaining the corresponding y_i and then deciding on replacement continues until all members of S have been considered. With the higher and higher number of generations points in S will come closer and closer. The stopping condition of DE, therefore, depends on the indication that the points in S have fallen into the region of attraction of the global minimum (or a minimizer). One way to measure this is to see if the absolute difference between the f_{max} and f_{min} falls below some given tolerance.

3. Differential Evolution Using Pre-calculated Differentials (DEPD)

DE proposed by Storn and Price generates differential vectors (one for each targeted point) in the mutation phase at each generation of the algorithm. Therefore, N differential vectors will be calculated by (3) in each generation. These differentials will gradually be shorter and shorter as the points in S become closer and closer. And this has two effects: (i) calculation of N differentials at each generation makes DE time consuming and (ii) it (calculation of the differential vectors at each generation) limits the exploratory feature of DE. The motivation of DEPD was to make DE faster and exploratory, and one way this could be achieved is to use pre-calculated differentials in the mutation operator. By simply making this change in DE the algorithm can be made substantially faster and exploratory. This change is implemented in the mutation phase.

We now explain how the pre-calculated differentials are used in DEPD. At the very first iteration of the algorithm all the differential vectors generated are kept in an array A. In the consecutive generations (say, the next M generations) when the point $x_i \in S$ are targetted, the intermediate point $\hat{x}_i$ is calculated using (3) by choosing a random point x_{r1} from S and a random differential from A. This process continues in the mutation phase for M generations of the algorithm before A is updated again with new differentials. Therefore, in DEPD mutation operation has two different modes: mode 1 where new differentials are used in (3) (mutation use three distinct vectors from S) and mode 2 where pre-calculated differentials stored in A are used in (3). Mode 1 of the mutation is switched on after every M generations, i.e., after every M generations the array A is updated with new differentials. The preceding considerations lead us to define the following DEPD algorithm.

The DEPD Algorithm

Step 1. Determine an initial set $S = \{x_1, x_2, \dots, x_N\}$ with points x_i, $i = 1, 2, \dots, N$ sampled randomly in Ω; evaluate $f(x)$ at each x_i, $i = 1, 2, \dots, N$. Take $N \gg n$, n being the dimension of the function $f(x)$. Set generation counter $k := 0$.

Step 2. Determine the points x_{max}, x_{min} and their function values f_{max}, f_{min} such that

$$f_{max} = \max_{x \in S} f(x) \quad \text{and} \quad f_{min} = \min_{x \in S} f(x).$$

Calculate the scaling factor F of the mutation operator using (4). If the stopping condition (e.g. $f_{\max} - f_{\min} < \varepsilon$) is satisfied, then stop.

Step 3. For each $x_i \in S$, determine y_i by the following two operations.

- Mutation: If $(k = 0)$ or $k \equiv 0 \pmod{M}$ execute mode 1 else mode 2.

  Mode 1: Randomly select three points x_{r1}, x_{r2} and x_{r3} from S except x_i, the running target and find the point $\hat{x}_i$ by the mutation rule (3) using the differential vector $(x_{r2} - x_{r3})$. If a component $\hat{x}_i^j$ falls outside Ω then it is found randomly in-between the jth lower and upper limits. Update the ith element of the array A with this differential.

  Mode 2: Randomly select a point x_{r1} from S and a differential vector from A and find the point $\hat{x}_i$ by the mutation rule (3). If a component $\hat{x}_i^j$ falls outside Ω then it is found randomly in-between the jth lower and upper limits.
- Crossover: Calculate the trial vector y_i corresponding to the target x_i from x_i and $\hat{x}_i$ using the crossover rule (5).

Step 4. Select each trial vector y_i for the $(k+1)$th generation using the accep tance criterion: replace $x_i \in S$ with y_i if $f(y_i) < f(x_i)$. Set $k := k + 1$ and go to Step 2.

We do not intend to present DEPD as a stand-alone algorithm for our current piece of work. We will use DEPD as an embedded component of the next algorithm whose motivations are as follows.

As it will be shown in posteriori that DEPD is much faster than DE, drawbacks remain. DEPD works for small size problems. Its stopping condition, the absolute difference between the $f_{\max}$ and $f_{\min}$ falling below some given tolerence, unnecessarily use a large number of function evaluations and its efficiency falls off as the number of dimension increases. Moreover, being a direct search method DEPD does not utilize any properties, for instance, the differentiability properties, of the function being optimized, even if such properties are available.

Although direct search-type algorithms for optimization have a role to play in certain situations where derivatives are not available it is quite clear that the use of gradient type techniques is generally preferable even when the gradients have to be computed numerically. Consequently we devise a new DE algorithm, the TDEPD algorithm, which can overcome the above mentioned drawbacks.

4. Topographical Differential Evolution Using Pre-calculated Differentials (TDEPD)

In this section we present our new algorithm, the TDEPD algorithm. TDEPD is designed for large-dimensional problems. The new algorithm fundamentally differs from DEPD in two ways. First it introduces an auxilary set S_a alongside S for exploration and second it uses topographical information on the objective function for efficient multistart. The effectiveness of the auxiliary set S_a for large-dimensional problems was shown in [16].

Except for the topographical phase, used for multistart, the difference between DEPD and TDEPD is the set S_a. In order to use as much information as possible of the points in S we introduce S_a of N points alongside S in DEPD. Initially, two sets each containing N points are generated in the following way; iteratively sample two points from Ω, the best point x_i going to S and the other x_i' to S_a ($f(x_i) < f(x_i')$). The process continues until each set has N points. The search process then updates both S and S_a simultaneously with generations. The reason for this is to make use of potential trial points which are normally rejected in DE/DEPD. At each generation, unlike DE/DEPD which updates one set, S, by the acceptance rule, TDEPD updates both sets S and S_a. In its acceptance phase, if the trial point y_i, corresponding to the target x_i, does not satisfy the greedy criterion $f(y_i) < f(x_i)$ then the point y_i is not abandoned altogether, rather it competes with its corresponding target x_i' in the set S_a. If $f(y_i) < f(x_i')$ then y_i replaces x_i' in the auxiliary set S_a. The potential points in S_a then can be used for further exploration and exploitation. The topographical phase utilizes topographical information on the objective function. Later we describe how TDEPD incorporates the topographical phase for multistart purpose but first we describe how the graph minima of a function are calculated in the topographical phase.

Our utilization of topograpical information is different from that of Törn and Viitanen [13] in that we do not intend to obtain the graph minima using the 'enough cover' of Ω. We use a small subset of N_g points from S for graph minima. A simplified description of how graph minima are found is the following: For each point in the sample of size N_g the g nearest neighbor points are determined ($g \ll N_g$). Those points for which all g neighbors are inferior points, i.e., the function values are larger, are the graph minima. The number of graph minima is dependent on the sample size N_g and g. We demonstrate this by using a renewed but simple example where $f(x)$, the function value at a point x, is simply the square of the distance from the point x to origin. We consider the function

$$f(x) = {x_1}^2 + {x_2}^2. \tag{6}$$

In Figure 1, there are $N_g = 15$ points numbered 1–15. Looking at 2 nearest neighbours ($g = 2$) we see that points $1, 5, 6, 9, 10$ and 14 are graph minima,

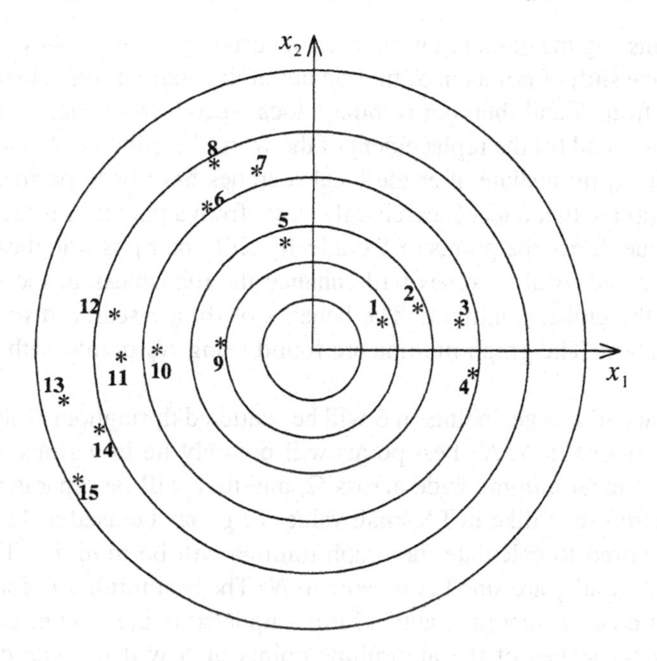

Figure 1. The contour plot of $f(x) = {x_1}^2 + {x_2}^2$ with $N_g = 15$ sample points.

for 3 points 1, 5 and 9 are graph minima, and for 4 nearest neighbours the graph minima are only at 1 and 9. The number of graph minima therefore depends on g and N_g points.

We now describe how the TDEPD works. At the very beginning, with the initial setting of the sets S and S_a TDEPD execute a generation by repeated use of mutation and crossover and updates both the sets using the acceptance rule. It then goes into the next generation and continues untill all initial points of the set S have shifted their positions (some may shift more than once). At this point TDEPD activates its first topographical phase and calculates the graph minima followed afterwards by multistart. The graph minima are obtained by using only N_g best points in S. In the multistart phase of TDEPD local searches are carried out from each of the graph minima. With the increase of generation from here, the second round of shifting process then continues again untill all points of S have been shifted again (before the second topographical phase an initial point in S must have shifted at least twice) before the next topographical phase and then multistart afterwards. This process of shifting, obtaining the graph minima and multistart continues repeatedly. Notice that in any consecutive two topographical or multistart phase all points of S are different. In other words, the topographical phase start with a complete new N_g points of S.

Since the DE/DEPD procedure gradually drives the N points in S towards the global minimizer two measures are introduced in TDEPD to lessen the chance of missing the global minimizer in the driving process. They are: (a) after a complete shift of points in S, finding out of the graph minima using the N_g best points from S and then performing a local search from each of the graph minima found, and (b) the replacement of the worst N_g points in S with the best N_g points in S_a immediate after the local searches have been performed. The benefits of (a) are that a local search only starts from a potential point with low function value. Since the points in S gradually shift their position these periodically scrutinized local searches will enhance the robustness of the algorithm in locating the global minimum. The benefits of (b) are search diversification and exploitation. The graph minima are found using N_g points with g nearest neighbours.

Clearly, at early stages points in S will be scattered throughout Ω and so will the best N_g points in S. N_g best points will possibly lie in various regions of attractions of local minima wide across Ω, and they will be separated by high regions. Therefore, unlike in TA small values of g may be used in TDEPD and the time required to calculate the graph minima will be minimal. This is because both N_g and g are small compared to N. The best minimum found in the local search phase is recorded and is further updated in the next phase of local search. In later stages of the algorithms points of S will become closer and closer, hopefully, around the global minimizer. Therefore, consecutive multistart may not produce a new better minimum. We use this as an indication to stop the TDEPD algorithm. In this way the TDEPD can avoid unnecessary function evaluation to form a dense cluster as in the case of DE/DEPD. In particular, if a consecutive number, say t, of local search phases does not produce any better minimum value than the previously found best minimum then the algorithm can be terminated. The step by step description of the new algorithm is as follows.

The TDEPD Algorithm

Step 1. Determine the initial sets $S = \{x_1, x_2, \ldots, x_N\}$ and $S_a = \{x'_1, x'_2, \ldots, x'_N\}$ with points sampled randomly in Ω. Initialize the generation counter k and the local phase counter t to zero. Initialize the Boolean array shift$[i]$ = false $\forall i = 1, 2, \ldots, N$.

Step 2. Determine the points $x_{\max}, x_{\min}$ and their function values $f_{\max}, f_{\min}$ such that

$$f_{\max} = \max_{x \in S} f(x) \quad \text{and} \quad f_{\min} = \min_{x \in S} f(x).$$

Calculate the scaling factor F of the mutation operator using (4). If the stopping condition, say $t \geqslant 4$, is satisfied then stop.

Step 3. For each $x_i \in S$, determine y_i by the following two operations:

- Mutation: If $(k = 0)$ or $k \equiv 0 \pmod{M}$ execute mode 1 else mode 2.

  Mode 1: Randomly select three points x_{r1}, x_{r2} and x_{r3} from S except x_i, the running target and find the point $\hat{x}_i$ by the mutation rule (3) using the differential vector $(x_{r2} - x_{r3})$. If a component $\hat{x}_i^j$ falls outside Ω then it is found randomly in-between the jth lower and upper limits. Update the ith element of the array A with this differential.

  Mode 2: Randomly select a point x_{r1} from S and a differential vector from A and find the point $\hat{x}_i$ by the mutation rule (3). If a component $\hat{x}_i^j$ falls outside Ω then it is found randomly in-between the jth lower and upper limits.
- Crossover: Calculate the trial vector y_i corresponding to the target x_i from x_i and $\hat{x}_i$ using the corssover rule (5).

Step 4. Update both the sets S and S_a for the next generation using the acceptance rule: If $f(y_i) < f(x_i)$ then do the following: (a) replace each $x_i \in S$ with y_i and (b) if shift$[i] \neq$ true then shift$[i] =$ true; otherwise replace $x_i' \in S_a$ with y_i if $f(y_i) < f(x_i')$. Set $k := k + 1$. If shift$[i] =$ true$\forall i = 1, 2, \ldots, N$ then go to Step 5, otherwise go to Step 3.

Step 5. Find the graph minima of the function, $f(x)$, using the best N_g points in S and perform a local search starting from each graph minimum. Keep a record of the very best minimum found so far, replace the worst N_g points in S with the best N_g in S_a. If the current phase of local minimization produces a better minimum than the current best minimum then set $t = 0$ otherwise set $t := t + 1$. Set shift$[i] =$ false$\forall i = 1, 2, \ldots, N$ and return to Step 2.

5. Numerical Results and Discussion

In this section the computational results are summarized. Computations were carried out using a SGI-Indy Irix-5.3. Random numbers were generated using the well tested procedure given in [17]. A limited memory BFGS algorithm (LBFGS) of Lui and Nocedal [18] was used as the local search algorithm. If two minimum values resulted from two different local search falls within 0.005 we consider them the same local minimum. Each of the DE algorithms was run 100 times on each of the test problems to determine the percentages of success in locating the global minimum. The performance is measured by criteria based on the number of function evaluation (FE), the cpu time (cpu) and the percentage of success. We even out the stochastic fluctuations in the number of function evaluations and cpu times by computing averages FE and cpu. We use $l_{\min} = 0.3$ and $N = 10n$ points in the population set S. These are within the suggested range [15] and are used in all implementations. The value of N_g in TDEPD should depend on the dimension of the problem. A choice of using the nearest integer N_g in $[0.25N, 0.3N]$ was found numerically sound. The effect of varying t and g is investigated.

5.1. Problems Used for Numerical Study

The performance of DE, DEPD and TDEPD were judged using easy, moderately difficult test problems and the hard LJ problem. The easy set A comprises of the Shekel family ($S5$, $S7$ and $S10$), the Hartmann family ($H3$ and $H6$) and the Goldstein and Price (GP) problem. These problems are well known to the global optimization community and are easily available in the literature. Therefore, we will not present any details of these problems. Besides these, we also consider another set, the set B of two relatively difficult test problems. These are 10-dimensional Rastrigin (RG) and Griewank (GW) functions. Finally we consider the scaled LJ potential function to test our new algorithm. These problems are given in Appendix I.

5.2. Numerical Results Using Test Problems

We begin our numerical studies by comparing DE and DEPD. To see the effect of pre-calculated differentials both algorithm were run using the same stopping rule. The stopping rule was chosen to satisfy $|f_{\max} - f_{\min}| \leqslant 10^{-4}$. We compare these two using a number of values for the parameter M. Notice that for $M = 0$ DE and DEPD are identical. For both the sets A and B successes are 100%. The same level of accuracy and the 100 per cent success conveniently allow us to compare the efficiency of the algorithms in terms of cpu and FE. We present the results for DEPD using $M = 1, 2, \ldots, 5$. Hence for a set of results for DE there are five sets for DEPD. We summarize these results in Table 1. We present the percentage of improvement by which DEPD outperforms DE. Improvements are recorded on total results. Total is the total of average cpu and FE for each function. Averages are taken over 100 runs. As predictable, Table 1 clearly shows the superiority of DEPD over DE in cpu. However, it is inferior to DE in FE for some cases. This superiority of DE over DEPD is by a very small margin and except for $M = 5$ this superiority can be ignored. For both the sets A and B superiority of DEPD in cpu increases with M but this trend is reversed at $M = 5$. Clearly the overall best results are obtained at $M = 4$ and for the rest of our numerical studies we use this value for M.

Our first objective was to justify the use of pre-calculated differential and with Table 1 we have been able to establish this. The pre-calculated differential component is embeded into the TDEPD algorithm. Therefore, next we compare only DE with TDEPD but before that we numerically study the effect of parameters in the new algorithm.

Table 1. Improvements of DEPD over DE

	M	1	2	3	4	5
A	cpu	11.89%	15.56%	15.34%	23.18%	14.47%
	FE	0.52%	−0.57%	−1.27%	−0.12%	−4.22%
B	cpu	10.73%	14.78%	15.56%	20.44%	13.1%
	FE	−0.75%	2.11%	−0.75%	3%	−2.46%

Parameter Selection. A motivation for the use of a small value of g has been discussed earlier. We have chosen $g = 2$ for all our implementations. For the easy set A the number function evaluations can be reduced by using a larger value for g. For instance using 3 or 4 nearest neighbours total FE can be reduced by about upto 4% while retaining the 100 per cent success rate. However, functions such as Griewank, Rastrigin and Lennard–Jones (LJ) with many local minima 2 nearest neighbours appeared to be a very good choice. A smaller value of g will also keep the time required in obtaining the graph minima minimal and it will also keep the algorithm robust in locating the global minimizer. For instance, if we use $N = 10n$, N_g = nearest integer to $0.3N$ and $g = 4$ and stop the algorithm if no new better minimizer is found after 4 consecutive multistart phases. We found that the algorithm fails thrice for the 10-dimensional Griewank function. However, for $g = 2$ or 3 there is not a single failure in 100 runs. This motivate us again to use the value of g as small as two.

One of the features of the new algorithm is that each time the graph minima have to be found after the shifting process has been completed. However, it may happen that a point is in the vicinity of a low lying local minimizer. In this situation, the algorithm will not perform the graph minima phase until this particular point has been shifted. This may take long time and the algorithm may loose its exploratory feature by delaying the multistart phase. Under these circumstances, we have adopted the following policy. Rather than the complete shift of all points we calculate the graph minima and the perform local search when 80% to 90% points in S have shifted their positions. However, this has the drawback of a current graph minimum being re-occurred in the next graph minima phase. We have made an experiment to study this using the sets A and B. We fix N_g, g and t as before and run the algorithm using 75%, 80%, 85% and 90% shifting. We have found that for 90% shifting the effect of repetitions were very negligible in the overall results. Moreover, failures (for Griewank) have occured for 75% and 80% shifting. Therefore, we use 90% shifting throughout the numerical calculations.

One of the difficulties of the global optimization is to decide when to stop. Here we have adopted a very simple but realistic strategy: stop when t number of multistart phase cannot produce a better minimizer. Again we determine an optimal value for t using an experiment. We have fixed N_g, g and use the 90% shifting policy. The algorithm was then run using $t = 2, 3, 4$ and 5. For the set A the algorithm fails four times (twice for each $H6$ and $S10$) for $t = 2$. For the rest of the values for t successes were 100%, but larger t means larger FE. For the set B, the algorithm fails 9 times (5 for GW and 4 for RG) for $t = 2$; it fails twice (GW) for $t = 3$. For the other values of t it was successful. Therefore, for all implementation we have used $t = 4$.

Numerical Comparison. Using the above parameter values we have run the TDEPD algorithm 100 times with 100% success on both the sets. The average results per run are presented in Table 2. In Table 2 we use the following notation: fe is the number of function evaluations required by the algorithm, fl is the number of function evaluations required by the local search algorithm, sf is the number of times the global minimum was located by the local search, lc is the number of local search performed per successful run. The data under the columns lc and sf are rounded to the nearest integer. Notice that FE is the sum of fe and fl. From Table 2 it is clear that the number of function evaluations (fe) needed by the algorithm is much higher than that of (fl) by local search. Hence our choice of a small g value was not detrimental to the behaviour of the algorithm in terms of FE. For the test set A the total number of local searches performed was 87 and 52 of them resulted in global minimizer, i.e., about 60% of the local searches successfully obtained the global minimum. For GP and H3 all the local searches found the global minimum. For the set B, on average, only 18% of the local searches were successful. The less successful case was the Griewank function with 3 successes out of 37 local searches. This is because many local minima are clustered around the global minimum for this function.

To see the good behaviour of the new algorithm over DE we now compare them using total FE and cpu. We first present the results obtained by DE. The values of the pair (FE, cpu) are (33 559, 1.45) and (396 268, 27.34) respectively

Table 2. Summarised results on sets A and B

	f	fe	fl	lc	sf	cpu
A	GP	1008	585	7	1	0.02
	S5	4190	161	15	10	0.14
	S7	3451	163	15	6	0.14
	S10	3328	161	15	6	0.15
	H3	909	172	12	12	0.10
	H6	3578	501	23	11	0.53
B	GW	6095	1561	37	3	1.48
	RG	100551	866	90	21	5.72

for A and B. For TDEPD these results are (18208, 1.08) and (109073, 7.2) respectively for A and B. Therefore, TDEPD outperforms DE on A by about 46% in FE and 26% in cpu, and on B by about 72% in FE and 74% in cpu, a considerable improvement. To test these algorithms further we consider higher-dimensional Griewank and Rastrigin functions, in particular we use fifteen and twenty dimensions of these problems. We again run the DE and TDEPD algorithms 100 times on each of these problems. As before, both the algorithms have retained 100 per cent success in obtaining the global minimum. In Table 3 we summarise the results. The least improvement by TDEPD is 42% and the best by 96%. Besides, because of the gradient based local search effect the accuracies of the final solutions for TDEPD are much better than that of DE. Table 3 further vindicate the superiority of the new algorithm for larger problems. This motivates us to test the TDEPD algorithm on the Lennard–Jones problems. This is presented in the next section.

5.3. Numerical Results Using the LJ Problem

We have chosen the scaled LJ potential function for further numerical studies. Details of this potential and its implementation are given in Appendix I. The minimization of this potential is difficult to achieve as it is a nonconvex optimization problem involving numerous local minima. We began this optimization using the DE algorithm. DE was able to locate the global minima for upto 4 particles in all of its 100 runs. For 5 atoms its success was only 70% and for 6 atoms the DE algorithm failed to locate the global minimum [19] completely. This problem seems to be exceptionally difficult, see Table 4. With further atoms added, the algorithm looses its success rate in locating the known global minima, and it became extremely slow and expensive in terms of FE. For instance, for seven and eight atoms it succeeded on avrage in about 11% ot the runs and the success of DE gradually dwindled with the increase of atoms. We then tested TDEPD on this problem and recorded the number

Table 3. Comparison using extended Griewank and Rastrigin

f	n	DE/TDEPD	FE	cpu	%FE	%cpu
RG	15	DE	1 182 795	119.32	42%	57%
		TDEPD	683 875	50.86		
	20	DE	6 940 320	771.17	49%	56%
		TDEPD	3 571 885	338.67		
GW	15	DE	235 455	22.33	96%	89%
		TDEPD	9876	2.39		
	20	DE	310 660	35.5	96%	89%
		TDEPD	12 153	3.77		

Table 4. Results of TDEPD on LJ potential

np	n	fe	fl	lc	lm	sf	f^*	ts
3	3	2524	858	33	15	17	−3.00	100
4	6	11 716	2574	96	55	33	−6.00	100
5	9	50 968	5076	179	114	50	−9.10	100
6	12	371 040	13 357	460	292	15	−12.71	6
7	15	1 630 464	18 990	395	275	70	−16.51	100
8	18	12 557 858	16 570	510	366	80	−19.82	100
9	21	54 942 699	17 342	568	390	58	−24.11	93
10	24	89 063 120	22 115	630	461	121	−28.42	95

of times the known global minimum [19] was located per run, the number of local searches performed and the number of local minima found per run. We summarise these results in Table 4. The results in Table 4 represent the average result of the runs for which the best minimum was successfully located. We use the following notation: np is the number of particles or atoms, lm is the number of different local minima found per successful run, lc is the number of local searches performed per successful run, ts is the percentage of successful runs out of 100 runs. The column under f^* represents the known best known minimum value [19]. Other notations are the same as in Table 2. From the table it can be seen that except for $np = 6$ the algorithm achieved close to 100% success rate. For $np = 6$ there are only 6 successes and the rest of the runs resulted in a near optimal value of -12.30. Table 4 also clearly shows that fe is considerably higher than fl. Since we have already used small g one way to increase fl is to use less shifting to start the topographical phase. To see the effect of this we run the algorithm again 100 times for atoms 3 to 10 and compared the results with that of 90% shifting presented in Table 4. We use 85%, 75% and 65% shifting policies and in the multistart phase we do not carry out local search from graph minima that have re-occured from the previous topographical phase. For each case the improvement of the algorithm on FE is significant. From higher to lower shifting fe decreases and fl increases gradually. At 75% the total fl nearly doubled and it begins to increase considerably at 65%. On the other hand the gradual decrease of fe for all cases is very high. We have noticed that the total successes (ts) for 85% and 75% remains more or less similar to that of 90% shifting (total ts are 694, 685 and 683 respectively for 90%, 85% and 75% shifting policies) However, at 65% the total ts was 658, i.e., about 5% lower than 694. We now compare FE used by the algorithm using 90% and 75% shifting. In total TDEPD for 75% shifting used 68% less function evaluations (FE) while retaining nearly the same level of success ts.

This motivate us to test the algorithm using 75% shifting on the test sets A and B. For the test set A the TDEPD algorithm using 75% shifting in total needed 59% less FE and 24% less cpu and and for the set B 75% shifting used 63% less FE and 38% less cpu. Therefore, the superiority of TDEPD over DE is much better than that reported in Table 3. However, for the Griewank function the algorithm failed 13 times. If we increase the shifting from 75% to 85% the algorithm fails twice for this function only. It therefore appears that for functions where the local minima are clustered around the global minimum TDEPD will need more shifting than for a function for which the minima are scatered.

6. Conclusion

We have developed a new global optimization algorithm. The algorithm proposes a new feature of differential evolution and incorporates topographical information for efficient multistart. We have shown how the differential evolution algorithm can be made faster using pre-calculated differentials. These features have made the algorithm very fast and robust.

In 100 independent runs the algorithm never failed to converge for both the test sets A and B. For the LJ potential problem the algorithm is tested for atoms upto 10. This problem has many local minima. Numerical experiences using this hard problem suggested that although the 90% shifting is workable but one can adjust this for less FE. The multistart feature strategically included in the algorithm has made it possible in locating the global minima. For problems with small number of local minima the algorithm will converge faster. This is because the inherent nature of the differential evolution algorithm (points move closer with the increasing number of iteration) and the choice of stopping condition: "no new better minimizer found" with points becoming closer. The value of t used in the stopping condition can be adjusted if the landscape of the underlying function is better understood.

A drawback of the algorithm is that it has several parameters. However, the algorithm was very robust in obtaining the global minimum even with various setting of these parameters. The only exception was that the value of g has to be small. The algorithm was very efficient and robust in locating the global minimum throughout all problems of small to moderately large dimensions. Consequently we feel that the algorithm could be used to solve a wide range of practical global optimization problems where the gradients are readily available. Research is continuing to develop an even more efficient and parameter free differential evolution algorithm.

Appendix I

Rastrigin (RG): $f(x) = 10n + \sum_{i=1}^{n}[x_i^2 - 10\cos(2\pi x_i)]$;
$x_i \in [-5.12, 5.12]$, $f(x) = 0$; $x^* = (0, 0, \ldots, 0)$.

Griewank (GW): $f(x) = 1 + \sum_{i=1}^{n} x_i^2/4000 - \prod_i^n \cos(x_i/\sqrt{i})$;
$x_i \in [-500, 500]$, $f(x) = 0$; $x^* = (0, 0, \ldots, 0)$.

Lennard–Jones (LJ): The minimization of the potential energy of N atoms can be written as

$$\min_x f(x) = \sum_{i=2}^{N} \sum_{j=1}^{i-1} v(r_{ij}),$$

where the pair potential is given by

$$v(r_{ij}) = r_{ij}^{-12} - 2r_{ij}^{-6}$$

and r_{ij} being the Euclidean interatomic distance between atom i and atom j. In this minimization atomic coordinates in the space are considered to be the independent variables. Therefore for N atoms there would normally be $n = 3N$ variables. However, we use the following argument to reduce the dimension of the problem to $n = 3N - 6$.

We first fix a particle at the origin and choose our second atom to lie on the positive x-axis. The third atom is chosen to lie in the upper half of the x-axis. Since the position of the first particle is always fixed and the second particle is restricted to the positive x-axis, this gives a minimization problem involving three variables for three atoms. For four atoms, additionally three variables (the Cartesian coordinates of the 4th atoms) are required to give a minimization problem in six independent variables. For each further particle, three variables (cordinates of the position of the particle) are added to determine the energetic of clusters. We have also defined the limits to the variables and these are given below.

The first and the third variables are taken to lie in $[0, 3]$. The second variable is taken to lie in $[-3, 3]$. The cordinates of the 4th atom taken to lie in $[-l_i, u_i]$ with $l_i = u_i = 3$ and then for next extra two atoms (5th and 6th) 6 variables involved are taken to lie in $[-l_i', u_i']$, where $l_i'(= u_i') = l_i(= u_i) + 1$. Similarly the next two atoms variables are taken to lie in $[-l_i'', u_i'']$ where $l_i''(= u_i'') = l_i'(u_i') + 1$ and then for each extra two particles the same process continues. Therefore, as an example, the six variables for the 5th and 6th particles lie in $[-4, 4]$ and for 9th and 10th the associated six variables lie in $[-6, 6]$.

Bibliography

[1] Ratschek, H. and Rokne, J.: *New Computer Methods for Global Optimization*, Ellis Horwood, Chichester, 1988.

[2] Törn, A. and Žilinskas, A.: *Global Optimization*, Springer-Verlag, Berlin, 1989.

[3] Horst, R. and Tuy, H.: *Global Optimization (Deterministic Approaches)*, Springer-Verlag, Berlin, 1990.

[4] Wood, G. R.: Multidimensional bisection and global optimization, *Comput. Math. Appl.* **21** (1991), 161–172.

[5] Floudas, A. and Pardalos, M.: *Recent Advances in Global Optimization*, Princeton University Press, Princeton, NJ, 1992.

[6] Goldberg, D.: *Genetic Algorithms in Search, Optimization and Machine Learning*, Addison-Wesley, Reading, MA, 1989.

[7] Ali, M. M., Törn, A. and Viitanen, S.: A numerical comparison of some modified controlled random search algorithms, *J. Global Optim.* **11** (1997), 377–385.

[8] Storn, R. and Price, K.: Differential evolution – a simple and efficient heuristic for global optimization over continuous spaces, *J. Global Optim.* **11** (1997), 341–359.

[9] Dekkers, A. and Aarts, E.: Global optimization and simulated annealing, *Math. Programming* **50** (1991), 367–393.

[10] Ali., M. M. and Storey, C.: Aspiration based simulated annealing algorithms, *J. Global Optim.* **11** (1997), 181–191.

[11] Ali, M. M., Storey, C. and Törn A.: Applications of stochastic global optimization algorithms to practical problems, *J. Optim. Theory Appl.* **95** (1997), 545–563.

[12] Rinnoy Kan, A. H. G. and Timmer, G. T.: Stochastic global optimization methods; Part II: Multilevel methods, *Math. Programming* **39** (1987), 57–78.

[13] Törn, A. and Viitanen, S.: Topographical global optimization, In: A. Floudas and M. Pardalos (eds), *Recent Advances in Global Optimization*, Princeton University Press, Princeton, NJ, 1992.

[14] Ali, M. M. and Storey, C.: Topographical multilevel single linkage, *J. Global Optim.* **5** (1994), 349–358.

[15] Ali, M. M. and Törn, A.: Evolution based global optimization techniques and the controlled random search algorithm: Proposed modifications and numerical studies, submitted to the *J. Global Optim.*

[16] Ali, M. M. and Törn, A.: Optimization of carbon and silicon cluster geometry for Tersoff potential using differential evolution, In: A. Floudas and M. Pardalos (eds), *Optimization in Computational Chemistry and Molecular Biology: Local and Global Approaches*, Kluwer Academic Publishers, Dordrecht, 2000.

[17] Tezuka, S. and L'Ecuyer, P.: Efficient and portable combined Tausworthe random number generators, *ACM Trans. Modelling and Computer Simulation* **1** (1991), 99–112.

[18] Lui, D. C. and Nocedal, J.: On the limited memory BFGS method for large scale optimization, *Math. Programming* **45** (1989), 503–528.

[19] Leary, R. H.: Global optima of Lennard–Jones clusters, *J. Global Optim.* **11** (1997), 35–53.

Chapter 2

OPTIMAL TAX DEPRECIATION IN STOCHASTIC INVESTMENT MODEL*

V. I. Arkin
Central Economics and Mathematics Institute
Russian Academy of Sciences
Moscow, Russia

A. D. Slastnikov
Central Economics and Mathematics Institute
Russian Academy of Sciences
Moscow, Russia

Abstract The paper proposes a model of tax stimulation for investing a creation of new industrial enterprises under uncertainty. The state which is interested in realization of a certain investment project uses accelerated depreciation mechanism in order to attract an investor. A behavior of a potential investor is assumed rational. It means that while looking at economic environment investor chooses such an investment moment that maximizes expected net present value which depends on depreciation policy. Specific features of the model are irreversibility of initial investment and stochastic flow of future enterprise's profits. Authors find such a depreciation policy which maximizes expected tax payments from the project after investment. For actual depreciation methods (straight-line and declining-balance) it is obtained optimal depreciation rates which satisfy certain restrictions.

Keywords: Investment under uncertainty, income tax, tax depreciation, optimal depreciation

*This work is supported by RFH (projects 01–02–00415, 01–02–00430a), RFBR (project 00–01–00194), and INTAS (project 99–01–317).

G. Dzemyda et al. (eds.), Stochastic and Global Optimization, 19–32.
© 2002 *Kluwer Academic Publishers.*

1. Introduction

In the paper we propose a model of tax incentives optimization for investment projects with a help of the mechanism of accelerated depreciation. Unlike the tax holidays which influence on effective income tax rate, accelerated depreciation affects on taxable income.

In modern economic practice the state actively use for an attraction of investment into the creation of new enterprises such mechanisms as accelerated depreciation and tax holidays.

The problem under our consideration is the following. Assume that the state (region) is interested in realization of a certain investment project, for example, the creation of a new enterprise. In order to attract a potential investor the state decides to use a mechanism of accelerated tax depreciation. The following question arise. What is a reasonable principle for choosing depreciation rate? From the state's point of view the future investor's behavior will be rational. It means that while looking at economic environment the investor choose such a moment for investment which maximizes his expected net present value (NPV) from the given project. For this case both criteria and "investment rule" depend on proposed (by the state) depreciation policy. For the simplicity we will suppose that the purpose of the state for a given project is a maximization of a discounted tax payments into the budget from the enterprise after its creation. Of course, these payments depend on the moment of investor's entry and, therefore, on the depreciation policy established by the state.

We should note that tax privileges (in corporate income tax) issued from the depreciated policy are related to a given project, not personal investor. Thus, choosing a depreciation policy the state can combine (in principle) both fiscal and incentive functions of tax system. An investigation of the conditions for such a combination is one of the main purposes of the given paper.

Our approach is based on a model of the interaction between the state and investor under uncertainty. Similar model was proposed by authors in [1–3] for study of tax holidays mechanism. These models consist of three main parts: description of an investment project as a random cash flow embedded into actual tax system; description of the behavior of potential investor (investment waiting model); and description of the behavior of the state choosing by certain criteria parameters of the relevant tax incentives. Note that the model describing behavior of investor who chooses optimal timing moment while looking at random economic environment (geometric Brownian motion) was firstly proposed by McDonald and Siegel [4] (see also [5]).

The paper is organized as follows. Section 2 provides a formal description of the model of the interaction between the state and investor, gives the basic depreciation methods, formulates main assumptions. In Section 3 we investigate this model (Theorem 1) and give explicit dependencies of investor's NPV and expected tax payments on parameters of the model. The main result of this section (and of the whole paper) is a derivation of optimal depreciation policy which maximizes expected discounted tax payments into the budget.

2. Description of the Model

2.1. General Scheme

It will be considered a project of creation of a new enterprise (in production) in a certain region of a country as an object of investment.[1] Investments I, necessary for the project are considered to be instantaneous and irreversible, so that they can not withdrawn from the project and used for other purposes after the project was started (sunk costs).

One can think of an investment project as a certain sequence of costs and outputs expressed in units (the technological description of the project). For this reason while looking at the current prices on both input and output, the investor can evaluate the cash flow from the project. The most important feature of the model is the assumption that at any moment the investor can either *accept* the project and start with the investment, or *delay* the decision before obtaining new information on its environment (prices of the products, demand, etc.). For example, if someone wish to invest in the creation of a plant for fuel production, the prices of which increase, it makes sense to delay the investment in order to receive greater virtual profit (but not for too long because of the time discount effects).

Suppose that the project is started at the moment τ.

Let the gross income from the project at time t be $(x_t^\tau,\ t \geqslant \tau)$, and production cost at this moment is equal to $y_t^\tau + A_t^\tau$, where y_t^τ is material cost (including both wages and allowable taxes), and A_t^τ is depreciation charge. If γ denotes corporate income tax rate net after tax cash flow of investor at moment t equals

$$x_t^\tau - y_t^\tau - \gamma(x_t^\tau - y_t^\tau - A_t^\tau) = (1-\gamma)\pi_t^\tau + \gamma A_t^\tau,$$

where $\pi_t^\tau = x_t^\tau - y_t^\tau$.

Economic environment can be influenced by different stochastic factors (uncertainty in market prices, demand, etc.). For this reason we consider that the "profits" $(\pi_t^\tau, t \geqslant \tau, \tau \geqslant 0)$ are described by a family of random processes, defined in some probability space $(\Omega, \mathcal{F}, \mathbf{P})$ with the flow of σ-fields $(\mathcal{F}_t,\ \tau \geqslant 0)$. $\mathcal{F}_t$ can be considered as available information on the system up to the moment t, and random variables π_t^τ are supposed $\mathcal{F}_t$-measurable.

The expected present value of investor from the project discounted to the investment moment is written by the following formula

$$V_\tau = \mathbf{E}\left(\int_\tau^\infty [(1-\gamma)\pi_t^\tau + \gamma A_t^\tau]\mathrm{e}^{-\rho(t-\tau)}\,\mathrm{d}t \,\middle|\, \mathcal{F}_\tau\right), \tag{1}$$

where ρ is the discount rate, and the notation $\mathbf{E}(\cdot|\mathcal{F}_\tau)$ stands for the conditional expectation provided the information about the system until the moment τ.

At the same time we can calculate the tax payments into the budget that can be made by the project after investment. The expected tax payments from the firm into *the budget* discounted to the moment τ are equal to

$$T_\tau = \mathbf{E}\left(\int_\tau^\infty \gamma(\pi_t^\tau - A_t^\tau)\mathrm{e}^{-\rho(t-\tau)}\,\mathrm{d}t \,\middle|\, \mathcal{F}_\tau\right), \tag{2}$$

that is the tax on cumulative discounted taxable income from the project started at the moment τ.

2.2. Depreciation

We will represent the depreciation at moment t for the project launched at moment τ as $A_t^\tau = Ka_{t-\tau}$, where K is initial depreciation base, and $(a_t, t \geqslant 0)$ is the deterministic depreciation flow such that

$$a_t \geqslant 0, \qquad \int_0^\infty a_t\,\mathrm{d}t = 1. \tag{3}$$

This scheme contains various depreciation methods (in continuous time) which are used in reality. Some of them are described below.

1. Straight-line method (SL)
If L is a useful life of the asset, and $\lambda = 1/L$, then

$$a_t = \begin{cases} \lambda, & \text{when } 0 \leqslant t \leqslant L, \\ 0, & \text{when } t > L. \end{cases} \tag{4}$$

2. Declining-balance method (DB)
According to this method the flow of depreciation charges at any time is calculated as a fixed proportion of the net of amortization cost of assets. Thus at the moment of time t after the beginning of the assets exploration the following relation holds $Ka_t = \eta K_t$, where η is an instantaneous depreciation rate, and $K_t = K(1 - \int_0^t a_s\,\mathrm{d}s)$ is a book value of assets. This implies $a_t = \eta(1 - \int_0^t a_s\,\mathrm{d}s)$, therefore $\dot{a}_t = -\eta a_t$, i.e.

$$a_t = \eta\mathrm{e}^{-\eta t}. \tag{5}$$

For this method the book value of assets at the moment t is $K_t = K\mathrm{e}^{-\eta t}$ and the sum of depreciation charges at the interval $(t, t+\Delta)$ is equal to χK_t where $\chi = 1 - \mathrm{e}^{-\eta\Delta}$ does not depend on the moment t.

3. Sum-of-years-digits method
According this method assets are depreciated with linear decreasing rate. If L is a useful life of the asset then

$$a_t = \begin{cases} (L-t)/M, & \text{when } 0 \leqslant t \leqslant L, \\ 0, & \text{when } t > L, \end{cases} \tag{6}$$

where $M = \int_0^L t\,dt = L^2/2$ is an analogue of "sum of years digits" during the useful life for continuous time.

There are, of course, another methods embedding into above scheme. Such is, for example, a combination of DB and SL methods when for a certain period of time the depreciation charges are computed by declining-balance, and then the residual value of an asset is depreciated by straight-line method.

2.3. Purposes of the Participants

Active participants of the model, which pursue their own interests are the investor and the state.

The purpose of the investor is to find a moment for investment (the rule of investing), which depends on previous (but not future) observations of the environment, so that its net present value (NPV) will be maximal within given tax system, i.e.

$$\mathbf{E}(V_\tau - I)e^{-\rho\tau} \to \max_{\tau}, \tag{7}$$

where $\mathbf{E} = \mathbf{E}(\cdot|\mathcal{F}_0)$ is the sign of an expectation (provided the known data about the system at moment $t = 0$), and maximum is considered over all the "investment rules", i.e. moments τ, depending only on observations of the environment up to this moment (the Markovian moment, i.e. such that an event $\{\tau \leqslant t\}$ belongs to σ-algebra $\mathcal{F}_t$ for all t).

Let us denote the optimal moment for the investment (investment rule) in the problem (7) under given depreciation flow $D = (a_t, t \geqslant 0)$ as $\tau^*(D)$.

The state can have many motives for the attraction of the investor, other than the purely fiscal (such as increase in employment, improvements in infrastructure, etc.). However, in this paper we will focus on the fiscal interest of the state. Namely, knowing the optimal behavior of the investor (optimal investment rule in (7)), the state will try to choose such depreciation policy that maximizes discounted tax payments from this project into the budget over all D from a given set of available depreciation flows $\mathcal{D}$

$$\mathbf{E}\big(T_{\tau^*(D)}e^{-\eta\tau^*(D)}\big) \to \max_{D\in\mathcal{D}}, \tag{8}$$

where T_τ is determined by the formula (2).

Let D^* be optimal depreciation flow for this project, i.e. solution of the problem (8) (if it exists).

The interaction of the investor and the state can be interpreted from the game-theoretical point of view. One player (investor) chooses the optimal strategy (moment for investment) depending on the strategy of the other player (depreciation flow, established by the state), and the other player, who knows the optimal reply of the first one, chooses his own strategy which maximizes his gain. In this situation "the state" is the dominant player and the pair $(D^*, \tau^*(D^*))$ can be viewed as a Stackelberg equilibrium (see, for example, [6]) in the game "state–investor".

2.4. Main Assumptions

The amount of investment I is constant (in time).

We assume that initial depreciation base is a fixed share of total amount of investment, i.e. $K = \psi I$ where $0 \leqslant \psi \leqslant 1$.

Investor's cash flow at moment t after investment (started at moment τ) is described by the following stochastic equation:

$$\pi_t^\tau = \pi_\tau + \int_\tau^t \pi_s^\tau(\alpha\, ds + \sigma\, dw_s), \quad t \geqslant \tau, \tag{9}$$

where α and σ are real numbers ($\sigma \geqslant 0$), $(w_s,\ s \geqslant 0)$ is a standard Wiener process (Brownian motion), and the process $(\pi_t,\ t \geqslant 0)$ is specified by the equation

$$\pi_t = \pi_0 + \int_0^t \pi_s(\alpha\, ds + \sigma\, dw_s), \quad t \geqslant \tau,$$

with given initial state π_0, or, equivalently (by Ito formula),

$$\pi_t = \pi_0 \exp\left\{\left(\alpha - \frac{\sigma^2}{2}\right)t + \sigma w_t\right\}, \quad t \geqslant 0. \tag{10}$$

We consider that watching the current prices on both input and output, the investor can evaluate the initial cash flow from the project $\pi_\tau = \pi_\tau^\tau$ before investment will be made. So we will refer to the process $(\pi_t,\ t \geqslant 0)$ as "virtual" profits process. We assume also that the investor takes his decisions on the project observing the virtual profits. Thus it is natural to consider that σ-field $\mathcal{F}_t$ is generated by the values of the process π_s up to the moment t, i.e. $\mathcal{F}_t = \sigma(\pi_s,\ 0 \leqslant s \leqslant t)$. Knowing the information about virtual profits investor can calculate (see formula (1)) the present value from the project provided it would be started at that moment.

As one can see from formulas (9) and (10) both "virtual profits" process $(\pi_t,\ t \geqslant 0)$ and the conditional (regarded $\mathcal{F}_t$) distribution of "real profits" π_t^τ are geometric Brownian motion with parameters (α, σ). The hypothesis of geometric Brownian motion is typical for many financial models (in particular, real options theory). As it was noted (see, e.g., [3]) such a hypothesis follows from a general assumptions about stochastic process (independence, homogeneity, continuity). The parameters of the geometric Brownian motion α and σ have a natural economic interpretation, namely, α is an expected instantaneous rate of profits change; σ^2 is an instantaneous variance of rate of profits change (volatility of the project).

We should emphasize, that the expected rate of profits change does not have to be positive. Negative α means that profits flow decreases with time (on average); and when $\alpha = 0$ it changes around the initial mean.

Now we can write the explicit formulas for investor's Present Value (1) and tax payments into the budget (2). We have

$$\mathbf{E}(\pi_t^\tau | \mathcal{F}_\tau) = \pi_\tau e^{\alpha(t-\tau)}, \quad t \geqslant \tau.$$

$$\begin{aligned} V_\tau &= (1-\gamma) \int_\tau^\infty \mathbf{E}(\pi_t^\tau | \mathcal{F}_\tau) e^{-\rho(t-\tau)}\, dt + \gamma K \int_\tau^\infty a_{t-\tau} e^{-\rho(t-\tau)}\, dt \\ &= (1-\gamma)\pi_\tau \int_0^\infty e^{-(\rho-\alpha)t}\, dt + \gamma K \int_0^\infty a_t e^{-\rho t}\, dt \\ &= \pi_\tau \frac{1-\gamma}{\rho-\alpha} + \gamma K A, \quad \text{where } A = \int_0^\infty a_t e^{-\rho t}\, dt; \end{aligned} \tag{11}$$

$$\begin{aligned} T_\tau &= \gamma \int_\tau^\infty (\mathbf{E}(\pi_t^\tau | \mathcal{F}_\tau) - A_t^\tau) e^{-\rho(t-\tau)}\, dt \\ &= \gamma \int_\tau^\infty (\pi_\tau e^{\alpha(t-\tau)} - K a_{t-\tau}) e^{-\rho(t-\tau)}\, dt \\ &= \gamma \left(\frac{\pi_\tau}{\rho-\alpha} - KA \right). \end{aligned} \tag{12}$$

From these formulas one can see that both Present Value of the investor and cumulative tax payments into the budget which define the criteria in behavior of the participants (7) and (8) depend only on integral discounted depreciation A (not on the whole depreciation flow D).

3. Investigation of the Model

In this section we provide a solution of the model formulated above. As it turned out, it can be obtained in an explicit (analytical) form.

3.1. Solution of the Investor Problem

The problem which the investor faces is an optimal stopping problem for the stochastic process. The relevant theory is well developed (see, for example, [7]), but there are very few problems which have a solution in an explicit form, and problem (7) belongs to this type.

Let β be a positive root of the quadratic equation

$$\frac{1}{2}\sigma^2 \beta(\beta-1) + \alpha\beta - \rho = 0. \tag{13}$$

We should point out that $\beta > 1$ whenever $\rho > \alpha$. For the deterministic case $\sigma = 0$, we have $\beta = \rho/\alpha$ whenever $\alpha > 0$, and there is no positive root of Equation (13) whenever $\alpha \leqslant 0$, but we find it convenient to consider $\beta = \infty$.

Let us denote $k = \beta/(\beta - 1)$.

THEOREM 1. *Let the virtual profits from the project* $(\pi_t,\ t \geqslant 0)$ *evolved according to the geometric Brownian motion* (10), *and* $\rho > \alpha$.

Then the optimal moment for the investment is

$$\tau^*(D) = \min\{t \geqslant 0 : \pi_t \geqslant \pi^*(D)\}^2, \tag{14}$$

where $\pi^*(D) = (1 - \gamma\psi A)\tilde{\pi}$, $\tilde{\pi} = k\frac{\rho-\alpha}{1-\gamma}I$, $A = \int_0^\infty a_t e^{-\rho t}\, dt$.

This theorem shows that the optimal moment for the investment begins when the virtual profit achieves a critical level $\pi^*(D)$. Formulas of this type (for the case of geometric Brownian motion) are given in [4,5] (for a more simple model of the investor) and for a close model in [1,3].

One can look at formula (14) from another point of view.

Using equality (11) for the investor's Present Value V_τ one can see that

$$V_\tau - I = \frac{1-\gamma}{\rho - \alpha}\pi_\tau - (1 - \gamma\psi A)I.$$

Applying known results on the optimal stopping for geometric Brownian motion we have that the optimal moment for the investment equals to the first moment when the Profitability Index for the project V_τ/I achieves threshold level $q = k(1 - \gamma\psi A) + \gamma\psi A$ (which is greater than one). Detailed analysis of this phenomenon and its connection with the known Jorgenson rules and Tobin's ratio q can be found in [5].

In order to avoid trivial moment of investment $\tau^*(D) = 0$, we will further suppose that the initial value of the virtual profit π_0 satisfies $\pi_0 < \pi^*(D)$.

The optimal moment for the investment $\tau^*(D)$ is not always finite. If parameters of the process are such that $\alpha < \sigma^2/2$, then with the positive probability $1 - \exp\{(1 - 2\alpha/\sigma^2)\log(\pi_0/\pi^*(D))\}$, the project can remain non-invested; if the expected rate of profits change is large enough regarded its volatility, $\alpha \geqslant \sigma^2/2$, then the project will be invested in (with the probability one)[3].

If we know the optimal moment for the investment, we can find the expected optimal net income for the investor as well as the relevant expected tax payments from the project into the budget. Let us denote the expected discounted net income of the investor under the condition of optimal behavior (i.e. maximal value of the function in (7)) as $\mathcal{V}(D)$, and let $\mathcal{T}(D) = \mathbf{E}(T_{\tau^*(D)}e^{-\rho\tau^*(D)})$ be the discounted tax payments into the budget under the optimal behavior.

THEOREM 2. *Let the virtual profits from the project* $(\pi_t, t \geqslant 0)$ *be a process of geometric Brownian motion* (10), *and* $\rho > \alpha$. *Then, the following formulas hold:*

(1) $\mathcal{V}(D) = (1 - \gamma\psi A)^{-\beta+1}(k-1)\left(\frac{\pi_0}{\tilde{\pi}}\right)^{\beta} I,$

(2) $\mathcal{T}(D) = \gamma(1 - \gamma\psi A)^{-\beta}\left(\frac{\pi_0}{\tilde{\pi}}\right)^{\beta}\left(k\frac{1-\gamma\psi A}{1-\gamma} - \psi A\right)I,$

(3) $\mathbf{E}\tau^*(D) = \frac{1}{\alpha - \sigma^2/2}\log\frac{\pi^*(D)}{\pi_0} \quad \textit{if } \alpha > \frac{\sigma^2}{2},$

$\mathbf{E}\tau^*(D) = \infty \quad \textit{if } \alpha \leqslant \frac{\sigma^2}{2},$

where $\pi^*(D)$ *and* $\tilde{\pi}$ *are defined in Theorem* 1.

This theorem is proved analogously to [3, Theorem 2] using formulas (11)–(12).

3.2. Solution of the State Problem: Optimal Depreciation Flow

In this section we find an optimal depreciation flow $D = (a_t, t \geqslant 0)$ that provides maximal expected discounted tax payments from the project into the budget.

Let $\mathcal{D}$ be a given set of all available (for the state control) depreciation flows D, A be a discounted depreciation (as it is defined in (11)) after tax holidays

$$A = A(D) = \int_0^\infty a_t e^{-\rho t}\,dt,$$

and

$$\min_{D\in\mathcal{D}} A(D) = \underline{A}, \qquad \max_{D\in\mathcal{D}} A(D) = \bar{A}.$$

We assume that the set of available depreciation flows is enough "rich" in the following sense.

(C1) For any value a, $\underline{A} < a < \bar{A}$ there exists depreciation flow $D \in \mathcal{D}$ such that $A(D) = a$.

In other words, set $\{a : a = A(D),\ D \in \mathcal{D}\}$ will be the interval $[\underline{A}, \bar{A}]$.

As one can see from Theorem 2 the expected discounted tax payments from the project into the budget under optimal behavior of the investor can be written as follows

$$\mathcal{T}(D) = \gamma g(u)\left(\frac{\pi_0}{\tilde{\pi}}\right)^{\beta} I, \quad \text{where } u = \gamma\psi A,$$

$$g(u) = (1-u)^{-\beta}\left(k\frac{1-u}{1-\gamma} - \frac{u}{\gamma}\right). \tag{15}$$

Thus the problem of finding an optimal depreciation policy by the state (see Section 2.3)

$$\mathcal{T}(D) \to \max_{D \in \mathcal{D}},$$

is reduced (due to formula (14)) to a more simple problem of maximization of a function at some interval, namely,

$$g(u) \to \max_{\underline{u} \leqslant u \leqslant \bar{u}}, \tag{16}$$

where $\underline{u} = \gamma\psi\underline{A}$, $\quad \bar{u} = \gamma\psi\bar{A}$.

We have from relation (15)

$$\begin{aligned} g'(u) &= \beta(1-u)^{-\beta-1}\left(k\frac{1-u}{1-\gamma} - \frac{u}{\gamma}\right) - (1-u)^{-\beta}\left(\frac{k}{1-\gamma} + \frac{1}{\gamma}\right) \\ &= (1-u)^{-\beta-1}G(u), \end{aligned} \tag{17}$$

$$\begin{aligned} G(u) &= \beta\left(k\frac{1-u}{1-\gamma} - \frac{u}{\gamma}\right) - (1-u)\left(\frac{k}{1-\gamma} + \frac{1}{\gamma}\right) \\ &= (\beta-1)k\frac{1-u}{1-\gamma} - \beta\frac{u}{\gamma} - \frac{1-u}{\gamma} \\ &= \beta\left(\frac{1-u}{1-\gamma} - \frac{u}{\gamma}\right) - \frac{1-u}{\gamma} = \beta\frac{\gamma-u}{\gamma(1-\gamma)} - \frac{1-u}{\gamma}. \end{aligned} \tag{18}$$

$G(u)$ is decreasing function since $G'(u) = -\beta/(1-\gamma) - (\beta-1)/\gamma < 0$.

In order to avoid "degenerative" cases we assume that the following requirement holds.

(C2) $\psi\bar{A} < 1$.

Note that this condition implies $\gamma > u$ for $u \leqslant \bar{u}$.

Let us pass to solving the maximization problem (16). Let u^* denote an optimal solution in (16).

If $G(\underline{u}) \leqslant 0$ then $G(u) \leqslant 0$ whenever $u > \underline{u}$, and due to (17) $g(u)$ decreases in u, therefore $u^* = \underline{u}$. Formula (18) and condition (C2) imply that $G(\underline{u}) \leqslant 0$ is equivalent to

$$\beta \leqslant \underline{\beta} = \frac{1-\gamma}{\gamma} \cdot \frac{1-\underline{u}}{1-\underline{u}/\gamma} = \frac{1-\gamma}{\gamma} \cdot \frac{1-\gamma\psi\underline{A}}{1-\psi\underline{A}}. \tag{19}$$

Similarly one can obtain that $G(u) > 0$ whenever $u \leqslant \bar{u}$ is equivalent to

$$\beta \geqslant \bar{\beta} = \frac{1-\gamma}{\gamma} \cdot \frac{1-\bar{u}}{1-\bar{u}/\gamma} = \frac{1-\gamma}{\gamma} \cdot \frac{1-\gamma\psi\bar{A}}{1-\psi\bar{A}}. \tag{20}$$

Hence, in this case $g(u)$ increases in u, and therefore, $u^* = \bar{u}$.

If $\underline{\beta} < \beta \leqslant \bar{\beta}$ then $g(u)$ attains maximum at the point u^* such that $G(u^*) = 0$, and

$$u^* = \frac{\beta\gamma - 1 + \gamma}{\beta - 1 + \gamma} = \gamma\left(\beta - \frac{1-\gamma}{\gamma}\right)/(\beta - 1 + \gamma). \tag{21}$$

Collect the above results we have the following

THEOREM 3. *Let conditions* (C1)–(C2) *hold. Then depreciation flow* $D^* = (a_t^*,\ t \geqslant 0)$ *is optimal if and only if the integral discounted depreciation* $A^* = \int_0^\infty a_t^* \mathrm{e}^{-\rho t}\,\mathrm{d}t$ *satisfies the following relations*

$$A^* = \begin{cases} \underline{A}, & \textit{when } \beta \leqslant \underline{\beta}, \\ \dfrac{\beta - (1-\gamma)/\gamma}{\psi(\beta - 1 + \gamma)}, & \textit{when } \underline{\beta} < \beta \leqslant \bar{\beta}, \\ \bar{A}, & \textit{when } \beta > \bar{\beta}, \end{cases}$$

where boundaries $\underline{\beta}$ *and* $\bar{\beta}$ *are defined in* (19) *and* (20) *respectively.*

Now we give the relevant results for the major classes of depreciation methods, specified in Section 2.2.

For the straight-line method with depreciation rate λ (see (4)) we have

$$A = A^{\mathrm{SL}}(\lambda) = \frac{\lambda}{\rho}\left(1 - \mathrm{e}^{-\rho/\lambda}\right).$$

Note that $A^{\mathrm{SL}}(\lambda)$ is increasing (in λ) function. We can also put $A^{\mathrm{SL}}(0) = 0$, $A^{\mathrm{SL}}(\infty) = 1$.

Assume that available depreciation rates have to place between two boundaries $\underline{\lambda}$ and $\bar{\lambda}$, i.e. $0 < \underline{\lambda} \leqslant \lambda \leqslant \bar{\lambda} < \infty$. As one can easy see the conditions (C1)–(C2) are valid for this case. Let us denote

$$\underline{\beta}^{\mathrm{SL}} = \frac{1-\gamma}{\gamma} \cdot \frac{1 - \gamma\psi A^{\mathrm{SL}}(\underline{\lambda})}{1 - \psi A^{\mathrm{SL}}(\underline{\lambda})}, \qquad \bar{\beta}^{\mathrm{SL}} = \frac{1-\gamma}{\gamma} \cdot \frac{1 - \gamma\psi A^{\mathrm{SL}}(\bar{\lambda})}{1 - \psi A^{\mathrm{SL}}(\bar{\lambda})}.$$

COROLLARY 1. *An optimal depreciation rate* λ *for SL method under restrictions* $\underline{\lambda} \leqslant \lambda \leqslant \bar{\lambda}$ *has the following form*

$$\lambda^* = \begin{cases} \underline{\lambda}, & \textit{when } \beta \leqslant \underline{\beta}^{\mathrm{SL}}, \\ \tilde{\lambda}, & \textit{when } \underline{\beta}^{\mathrm{SL}} < \beta \leqslant \bar{\beta}^{\mathrm{SL}}, \\ \bar{\lambda}, & \textit{when } \beta > \bar{\beta}^{\mathrm{SL}}, \end{cases}$$

where $\tilde{\lambda}$ *is a root of the equation* $\psi A^{\mathrm{SL}}(\lambda) = (\beta - \frac{1-\gamma}{\gamma})/(\beta - 1 + \gamma)$.

For the declining-balance method described in Section 2.2 (see formula (5)) we have

$$A = A^{\mathrm{DB}}(\eta) = \frac{\eta}{\rho + \eta}.$$

The function $A^{\mathrm{DB}}(\eta)$ also increases (in η).

If there are certain restrictions on depreciation rate $\underline{\eta} \leqslant \eta \leqslant \bar{\eta}$, then (C1)–(C2) are also satisfy. Similar to previous consideration let us denote

$$\underline{\beta}^{\mathrm{DB}} = \frac{1-\gamma}{\gamma} \cdot \frac{1-\gamma\psi A^{\mathrm{DB}}(\underline{\eta})}{1-\psi A^{\mathrm{DB}}(\underline{\eta})}, \qquad \bar{\beta}^{\mathrm{DB}} = \frac{1-\gamma}{\gamma} \cdot \frac{1-\gamma\psi A^{\mathrm{DB}}(\bar{\eta})}{1-\psi A^{\mathrm{DB}}(\bar{\eta})}.$$

COROLLARY 2. *An optimal depreciation rate η for DB method under restrictions $\underline{\eta} \leqslant \eta \leqslant \bar{\eta}$ is the following*

$$\eta^* = \begin{cases} \underline{\eta}, & \text{when } \beta \leqslant \underline{\beta}^{\mathrm{DB}}, \\ \rho b^*/(1-b^*), & \text{when } \underline{\beta}^{\mathrm{DB}} < \beta \leqslant \bar{\beta}^{\mathrm{DB}}, \\ \bar{\eta}, & \text{when } \beta > \bar{\beta}^{\mathrm{DB}}, \end{cases}$$

where $b^* = \frac{1}{\psi}(\beta - \frac{1-\gamma}{\gamma})/(\beta - 1 + \gamma)$.

4. Some Numeric Examples

In this section we calculate optimal depreciation rates in some examples using adjusted real data. As reasonable sets of the parameter values, we will consider (similar as in [3,8]) investment projects with expected growth rate varying from -1% to 3%, volatility $\sigma \sim 0.1 \div 0.2$, and discount rate ρ from 10% to 20% (all parameters here and further will be annual). The corporate income tax rate γ is equal to 30% (as in Russia now).

In the tables below we provide some values of optimal depreciation rates for SL and DB methods under various combinations of parameters α, σ, ψ and ρ. For the simplicity we use Corollaries 1 and 2 with trivial restrictions: $\underline{\lambda} = \underline{\eta} = 0$, $\bar{\lambda} = \bar{\eta} = \infty$. So, λ^* will denote the optimal SL depreciation rate from Corollary 1. As for DB method we shall consider *annual* optimal depreciation rate $\chi^* = 1 - e^{-\eta^*}$ where η^* is taken from Corollary 2. As we pointed out in Section 2.2, this rate can be considered as a share of a book value of an assets which has been depreciating during a unit period of time (year).

Table 1 shows the dependence of the annual optimal depreciation rates λ^* and χ^* on the expected growth rate in profits (α). Volatility of the project is taken as $\sigma = 0.15$, discount rate is equal to 15%, and share of initial depreciation base in total amount of investment $\psi = 0.8$.

Table 1.

Exp. rate of growth α	Optimal SL λ^*	Optimal DB χ^*
−1%	0.24	0.35
0%	0.17	0.26
1%	0.12	0.18
2%	0.08	0.12
3%	0.05	0.07

Table 2.

Share of deprec. assets ψ	Optimal SL λ^*	Optimal DB χ^*
1	0.10	0.16
0.9	0.13	0.19
0.8	0.17	0.26
0.7	0.26	0.38

Table 3.

Discount rate ρ	Optimal SL λ^*	Optimal DB χ^*
($\alpha = 0\%$)		
10%	0.07	0.11
15%	0.17	0.26
20%	0.32	0.44
($\alpha = 3\%$)		
10%	0	0
15%	0.05	0.07
20%	0.13	0.19

As one can see from this table, when expected growth rate in profits increases, optimal depreciation rate decreases.

Table 2 illustrates the dependence of the same indicators on the share of initial depreciation base in total amount of investment ψ. We take the project with parameters $\alpha = 0$, $\sigma = 0.15$ and discount – 15%.

Increasing of optimal depreciation when ψ decreases is a simple consequence of Corollaries 1 and 2. Let us remember that we consider the case when upper boundaries $\bar{\lambda}$, $\bar{\eta}$ for depreciation rates are absent (or, equal to infinity).

Table 3 demonstrates the dependence of the above mentioned indicators both on the discount ρ and expected growth rate in profits α. As in previous tables we put volatility $\sigma = 0.15$ and share $\psi = 0.8$.

We should note that the majority of optimal annual DB depreciation rates χ^* are equal to about 1.5 times of corresponding SL depreciation rates, as it established in tax systems of Australia and, partially, France and Spain (the data are taken from [9]). Moreover, the calculation shows that optimal depreciation rates have enough reasonable values and correlated with statutory depreciation rates for Russia and OECD countries (see, e.g., [9]).

Notes

1. We will refer to a creation of a *new* enterprise, not the reconstruction of an *existing* one, because we mean that a *new* taxpayer will appear.
2. If a set of such t is empty, then put $\tau^*(D) = \infty$.
3. Similar formulas for a close model see, for example in [3, p. 23].

Bibliography

[1] Arkin, V. I. and Slastnikov, A. D.: Waiting to invest and tax exemptions, Working Paper WP/97/033, CEMI Russian Academy of Sciences, 1997.
[2] Arkin, V. I. and Slastnikov, A. D.: Tax incentives for investments into Russian economy, Working Paper WP/98/057, CEMI Russian Academy of Sciences, 1998.
[3] Arkin, V., Slastnikov, A. and Shevtsova, E.: Tax incentives for investment projects in the Russian economy, Working Papers Series No 99/03, EERC, Moscow, 1999.
[4] McDonald, R. and Siegel, D.: The value of waiting to invest, *Quart. J. Econom.* **CI**(4) (1986).
[5] Dixit, A. K. and Pindyck, R. S.: *Investment under Uncertainty*, Princeton University Press, Princeton, 1994.
[6] Aubin, J.-P.: *Nonlinear Analysis with Economic Applications*, Mir, Moscow, 1988.
[7] Shiryaev, A. N.: *Statistical Sequential Analysis*, Nauka, Moscow, 1969.
[8] McDonald, R.: Real options and rules of thumb in capital budgeting, In: M. J. Brennan and L. Trigeorgis (eds), *Project Flexability, Agency, and Competition (New Developments in the Theory and Application of Real Options)*, Oxford University Press, 1999.
[9] Cummins, J. G., Hassett, K. A. and Hubbard, R. G.: Tax reforms and investment: A cross-country comparison, *J. Public Econom.* **62**(3) (1996).

Chapter 3

GLOBAL OPTIMISATION OF CHEMICAL PROCESS FLOWSHEETS

I. D. L. Bogle* and R. P. Byrne
Centre for Process Systems Engineering
Department of Chemical Engineering
University College London
Torrington Place
London WC1E 7JE, U.K.
d.bogle@ucl.ac.uk

Abstract Some chemical process design systems include a numerical optimisation capability and this is increasingly being demanded. The incorporation of interval based global optimisation in modular based systems which dominate the market is difficult because of the way modules are used as black boxes. In this paper a way of implementing global optimisation by recasting the models in a generic way is discussed. Two interval based algorithms are presented with results on two simple process optimisation problems giving an idea of the price that may need to be paid for the convenience of the modular systems. The two interval algorithms are based on reformulating the problem to be able to provide tighter estimates for the lower bounds on convex nonlinearities.

Keywords: Interval based global optimisation, chemical process flowsheets, interval algorithms

1. Introduction

As in all branches of Engineering, optimisation of new designs and of the operation of manufacturing plant is critical for a competitive business. In the field of Chemical Engineering this has become more clearly codified in the computational systems used for design and for management of operations. Most design systems now have the capability to optimise the design, almost always using gradient based methods, and the control systems on process plant have on-line optimisers, again using gradient based methods, which try and

*Author to whom correspondence should be addressed.

G. Dzemyda et al. (eds.), Stochastic and Global Optimization, 33–48.
© 2002 *Kluwer Academic Publishers.*

ensure that the operations are running in the most efficient manner. This capability is now used routinely.

The need for process optimisation is becoming more widespread as the manufacturing objectives become more complex. Financial efficiency, however defined, is no longer the only optimisation criterion used. Safety is considered to be the number one priority for any design of new plant or modification of existing plant. Now increasingly the environmental performance of the plant is becoming a key issue in its acceptance by the regulatory authorities for permission to operate. Of course this comes from increasing pressure from society in general.

This often results in a trade off between economic performance and environmental performance but both can be formulated as optimisation problems. The traditional objective functions for economic performance are the net present value of a plant or the internal rate of return on investment [1]. Objective functions for environmental performance are less well developed or tested but some such as the sustainability index [2] or the critical water mass which is a measure of the pollutants discharged [3] have been proposed and tested in process design systems (e.g., [4]).

Many other optimisation problems arise in Chemical Engineering: fitting data to models, thermophysical property prediction, predicting the conformation of molecules – particularly difficult with large molecules such as proteins, and in the development of new products which are optimal for a particular use. These developments will ensure that optimisation remains at the heart of Chemical Engineering. Kallrath [5] published an excellent review of the optimisation needs in the Chemical Industry from an industrial perspective.

One of the big issues is that the models are frequently non-convex. These non-convexities arise from combinations of nonlinear functions and from discontinuities in the describing equations and their derivatives. This of course means that the likelihood of finding a local minimum is very high. Since all the problems of any significance are large it becomes impossible to exploit specific mathematical features of a particular problem to avoid finding local minima and so a systematic means of finding genuinely global minima, particularly when there are significant differences between local and global minima, becomes very important.

The purpose of the paper is to discuss the issues of implementation of interval based global optimisation algorithms in chemical process optimisation, with particular emphasis on how this might be used within the Sequential Modular systems so popular within the industry. In the following section the types of systems commonly used in process design and optimisation, Sequential Modular and Equation Oriented, are discussed, followed by a simple example from the process industry which has local and global optima. The interval based algorithms used in this work are then outlined. To use interval based methods in modular flowsheeting systems the models must be modified which is discussed in Section 6 followed by how the optimisation is implemented. Finally some numerical results contrast the two approaches setting out the price that must be paid for the convenience of modular systems.

2. Process Modelling Systems in Chemical Engineering

Chemical manufacturing processes are made up of interconnected processing units that convert raw materials into valuable products. The conversion can be chemical, physical such as change of temperature or of pressure, to separate out components, or to mix streams. Recycle of unconverted raw materials is an essential part of the majority of large scale production facilities.

The computational systems for designing or analysing such plants have mostly been developed to mimic this type of structure. Individual modules or procedures have been developed for each unit and are connected together in an appropriate calculation order. The recycle is handled iteratively and there are computational strategies for handling specifications placed on streams other than inputs, such as a desired product quality. These systems are called Sequential Modular systems (see [6] for a review). More recently Equation Oriented systems have begun to appear commercially where the equation set is assembled and solved simultaneously (see [7] for a review). This second class has much greater flexibility in that a wider range of numerical codes can be used and the extension to dynamical systems is straightforward. However the natural structure of the problem is lost during the solution phase and this causes difficulties in trouble shooting when problems arise in finding the solution. Sequential Modular systems dominate the market.

The desire for globally optimal solutions exists with both types of system and there has recently been increasing interest in deterministic and stochastic global optimisation methods in the Chemical Engineering literature ([5,8–16]).

3. A Simple Process Optimisation Example

Blending of raw materials to make products is a very common optimisation problem in the Chemical industry. For example every refinery undertakes such an optimisation regularly to find the most efficient production conditions for as many as forty oil products. The Haverly Pooling problem given in Figure 1 as formulated by Quesada and Grossmann [17] is a very small example of such a blending problem. The problem is to blend four feeds into two products to minimise the total cost. Each feed has a fixed composition, x_A, and cost, c(£/kmol hr). Each product has a cost, required composition and flowrate, F. The feeds are mixed to produce products which satisfy the quality requirements using mixer and splitter units to represent the different blending tanks.

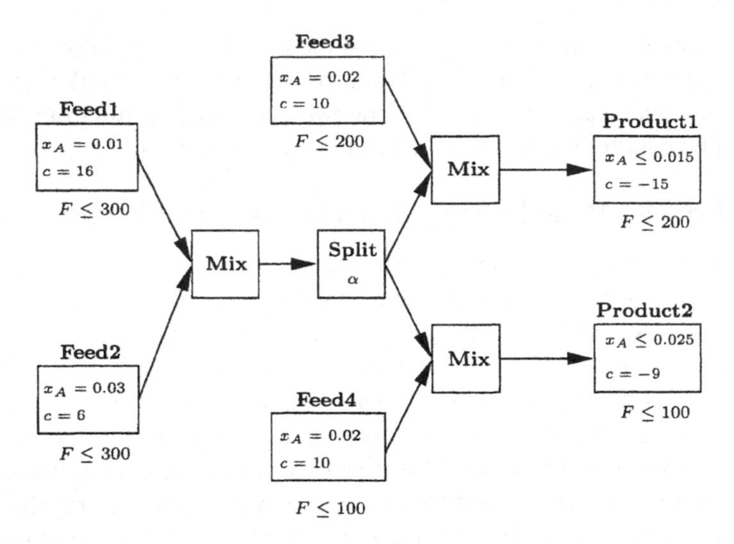

Figure 1. The haverly pooling problem.

The equation based formulation from [18] is

$$\begin{aligned} \min_{F, x_A} \quad & 6F_1 + 16F_2 + 10F_4 - 9F_5 + 10F_7 - 15F_8 \\ \text{s.t.} \quad & \\ & F_1 + F_2 - F_3 - F_6 = 0, \\ & F_3 + F_4 - F_5 = 0, \\ & F_6 + F_7 - F_8 = 0, \\ & 0.03F_1 + 0.01F_2 - F_3x_A - F_6x_A = 0, \\ & F_3x_A + 0.02F_4 - 0.025F_5 \leqslant 0, \\ & F_6x_A + 0.02F_7 - 0.015F_8 \leqslant 0, \end{aligned} \tag{1}$$

where x_A is the mole fraction of component A after mixing.

The problem can also be cast as a modular problem. All the mixers are the same, as are all the splitters. The feeds are unit operations without inputs where the flowrate is a parameter of the unit, and the product units represent quality requirements.

Given that this is one of many pooling/blending problems that may need to be solved it would be advantageous to be able to formulate the problem as a modular problem; placing units onto the flowsheet and 'drawing' the streams to connect them. It is natural to view the problem in terms of interconnected units with each unit performing some transformation of the input stream to provide the output stream.

This allows unit models to be reused in other flowsheets reducing the amount of work required to formulate the problem and reducing the scope for error in the formulation. It presents a view of the flowsheet which is the intuitive view for the Chemical Engineer.

The problem above is nonconvex so a local optimisation algorithm will not guarantee the global solution. Both Equation Oriented and Sequential Modular formulations require optimisation algorithms. However deterministic global optimisation algorithms cannot be applied in current modular flowsheeting systems and here we show how it can be done. We have chosen to use interval based methods and the algorithms are described in the next section.

4. Two Interval Algorithms

Interval methods have been in use for some years and there are two excellent texts available by Moore [19] and Hansen [20]. It is possible to significantly affect the efficiency of interval based methods by exploiting algebraic reformulation. The identification of convex terms in the objective function and constraints allows tighter interval bounds to be formulated. The number of variables that need to be branched on is restricted to those that appear in nonlinear terms and the lower bound produced is feasible with respect to any convex constraint. Two relaxations have been used in this work. The Natural Extension (NE) underestimates the convex nonlinear term by two linear functions which are based on the most conservative bounds for an interval arising from the nonlinear function. The Mean Value relaxation (MV) also produces two linear functions but based on the mean value theorem which in general provides a less conservative bound [21]. A combination of these two (MVNE) has also been used.

These ideas have been implemented in two depth-first branch and bound algorithms using interval analysis as the primary bounding method and using extra symbolic information where possible. The two variants are called Interval Global Optimisation Relaxation (IGOR) and Reduced Interval Global Optimisation Relaxation (RIGOR) because they use interval analysis and relaxation to efficiently solve constrained global optimisation problems [15]. An interval of a variable x is designated in upper case as X. Although the ideas are implemented in a depth-first algorithm there is no reason why they should not be applied in the classical interval breadth-first algorithm to locate all the global minimisers.

Interval analysis can be used to bound a very broad range of functions without any differentiability assumptions. The MV underestimators can be used to bound differentiable terms by reformulation of the interval bounding problem as a LP. The NE underestimators and the reformulations can be used when symbolic information is available.

4.1. The IGOR Algorithm

The IGOR algorithm is a depth-first branch and bound algorithm using interval inclusion and MV relaxation to construct a lower bounding LP. At each iteration the lowest box is removed from the list and bisected perpendicular to its longest edge.* Lower bounds are obtained for the new boxes and the boxes are returned to the list if the lower bounding problem is feasible.

4.1.1. Obtaining the Bounds.

For some box X^k the relaxed LP, P^k, is solved to obtain lower bounds. For problems of the form

$$\begin{aligned} \min_{x} \quad & f(x) + c^T x \\ \text{s.t.} \quad & g(x) + a^T x \leqslant 0, \\ & Ax - b \leqslant 0, \end{aligned} \tag{2}$$

where $f(x)$ and $g(x)$ are differentiable on X^k, then P^k is

$$\begin{aligned} \min_{x,\alpha,\beta} \quad & \alpha + c^T x \\ \text{s.t.} \quad & \max_i \{f_l(x, c_i)\} - \alpha \leqslant 0, \\ & \beta + a^T x \leqslant 0, \\ & \max_i \{g_l(x, c_i)\} - \beta \leqslant 0, \\ & Ax - b \leqslant 0, \end{aligned} \tag{3}$$

where $c_i \in \Re^n$ are the corners of X^k, $f_l(x, c_i)$ and $g_l(x, c_i)$ are the MV relaxations of $f(x)$ and $g(x)$ at c_i, the $\max(\ldots)$ terms are rewritten according to a smoothing formulation and bounds on the new variables α and β are obtained from the interval inclusions of $f(x)$ and $g(x)$ on X^k respectively. NE relaxations of $f(x)$ and $g(x)$ may also be added.

Any feasible solution to the LP problem solved above will provide a simply obtained candidate for the upper bound. This is the strategy used by the IGOR algorithm. In principle better alternatives could be found by solving a local optimisation problem but this was not found to improve the overall performance of the algorithm as in most cases the convergence of the algorithm is limited by the lower bounding problem.

4.2. The RIGOR Algorithm: Reduction

The RIGOR algorithm adopts the same approach as IGOR but an additional 'Reduce' step is added. Given an upper bound, $\overline{y}$, on the solution and a lower bounding LP, P^k, which we rewrite here as a LP with the extra variables incorporated into the vector x:

$$\begin{aligned} \min_{x} \quad & c^T x \\ \text{s.t.} \quad & Ax \leqslant b \end{aligned} \tag{4}$$

*The edges are scaled according to the widths of X_i^0.

the reduction step solves a set of LPs to tighten the bounds on each of the x variables. For each x_i that appears nonlinearly we solve

$$\begin{array}{ll} \min\limits_{x \in X^k} & x_i \\ \text{s.t.} & \\ & c^T x \leqslant \overline{y}, \\ & Ax \leqslant b \end{array} \tag{5}$$

to tighten the lower bound, $\underline{x_i}$, and solve

$$\begin{array}{ll} \max\limits_{x \in X^k} & x_i \\ \text{s.t.} & \\ & c^T x \leqslant \overline{y}, \\ & Ax \leqslant b \end{array} \tag{6}$$

to tighten the upper bound. If these problems are infeasible then the box, X^k can be deleted. There are many alternative reduction steps that could be used such as those used by the BARON algorithm [22] or in the approach of Hansen *et al.* [23]. The tightening steps used in RIGOR but not present in IGOR are especially useful when the initial bounds, X^k, are much larger than the feasible region described by the constraints.

RIGOR uses the best current upper bound, $\overline{y}$, and this is obtained using a local NLP algorithm before the first bisection. The NLP uses the center of X^0 as the starting point for the optimisation. For the rest of the iterations RIGOR uses the same method as IGOR.

5. Optimisation of Interval Modular Flowsheets

The following sections develop a modular flowsheeting approach which can be globally optimised and algorithms which can be applied to optimise such a flowsheet.

5.1. Extended Type Flowsheets

In order to formulate the modular flowsheeting problem to be solvable by deterministic global optimisation algorithms it is necessary to recast the problem in terms of extended type modules which work with extended type operations such as interval arithmetic.

An interval is a compound type made from two real numbers which is used to represent a range. An extended type is made from the compound type plus the rules that define operations on it. Thus, the interval compound used according to the rules of interval arithmetic is an extended type.

This is a specialised case of the more general ideas of Object Oriented Programming (OOP) where collections of data and functions (operators) together form objects, or types.

A number of extended arithmetic types are useful in process simulation:

- Intervals are an extended type for calculating the results of application of operators to ranges.
- Vectors and matrices are, in principle, extended types built from arrays of scalar types and rules to manipulate the data.
- The convex underestimators proposed by McCormick [24] provide the rules for adding, subtracting and multiplying convex underestimators. A suitable definition of the compound object which maintains the data required at each step makes this an extended type.
- The automatic differentiation (AD) rules proposed by Rall [25] are, when combined with a compound for the data, an extended type. Importantly the AD compound is built with arrays of some underlying type which can be a real number or an interval.

All of the operations provided by these extended types can be obtained with normal real arithmetic. The key to the encapsulation of data and operator rules is that the abstraction provided by the extended type reduces the complexity of the calculations and allows one abstraction to be built upon another.

Rather than construct many different flowsheets from units based on different types some form of model which can deal with many types generically is necessary. A generic model is a model that specifies the transformations that need to be applied to some underlying type, T, to obtain the output, as shown in Figure 2. The model needs to describe the operations and then the appropriate rules are applied for T. For example, a module which adds its inputs should use the interval arithmetic rules if the underlying type is an interval (T = interval) and the rules of AD if the underlying type is an AD type.

Once the types are constructed they can be used with the model to give the appropriate information. Because the generic model specifies *which* operations (or transformations) are to be applied and the type defines *how* those operations are applied to the compound, a generic model can be used with types that are unknown when the model is constructed. More details of the approach can be found in [26].

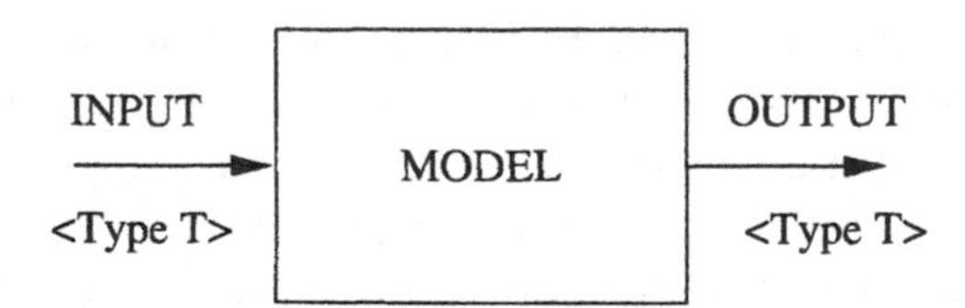

Figure 2. A generic model.

5.2. Interval Unit Operations

In order to solve the optimisation problem globally it is necessary to obtain upper and lower bounds on the problem. With genuinely black-box models it is only possible to evaluate the output at fixed points which provides no information about the behaviour of the model between those points. It is not possible to minimise the model rigorously if it is nonconvex.

If these models are rewritten as generic models using interval arithmetic so that the inputs are intervals and the outputs are intervals which bound the possible real outputs each unit model provides the information necessary to apply an interval global optimisation algorithm to the unit. The two very simple models, for mixers and splitters, required for the Haverly pooling problem are shown in Figure 3. Complex models can also be formulated in this way. Then, as with a normal modular flowsheet, it is possible to link unit operations with streams, which use intervals instead of real numbers, to get bounds on the outputs of the flowsheet.

This provides a means by which modular flowsheets can be constructed from unit operations based on interval analysis such that a global optimisation algorithm can be applied.

When a flowsheet is built from these general units using interval arithmetic the interval global optimisation algorithm can be applied. The algorithm is provided with initial bounds on the unit parameters which become the initial bound vector X^0. At iteration k, interval parameters, X^k, are chosen by the algorithm. Each unit calculates the output streams and the cost associated with the unit. The summation of these costs provides the objective function.

The design constraints on the products are added to the optimisation problem and the flowsheet is optimised to minimise the sum of the costs subject to these constraints.

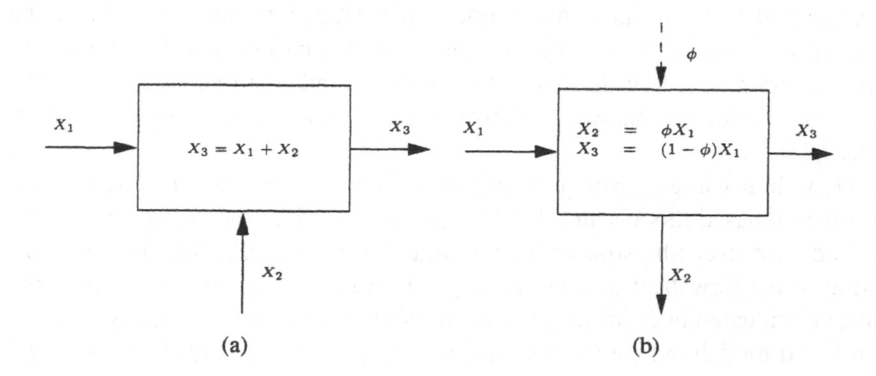

Figure 3. Unit operations based on interval arithmetic: (a) an interval mixer; (b) an interval splitter.

The interval flowsheet for the Haverly Poolong problem has been solved with a simple depth-first interval algorithm without reformulation. However with simple interval operations it does not converge to a unique solution in reasonable time. Examination of the list of possible solutions at this point indicates that the design constraints, which are nonconvex with respect to the independent variables, are not fully satisfied.

This is a problem in general for the application of interval optimisation when the solution lies on a nonlinear constraint which cannot be bounded tightly [27]. The problem is made worse because the evaluation through the flowsheet introduces a certain amount of redundant calculation which can make the bounds very conservative.

As noted in [27] the bounds can be improved by using a Mean Value extension of the constraints with optimal centres. This requires bounds on the gradient of the constraints. The gradient cannot be obtained by inspection because of the modular nature of the flowsheet. Perturbation does not provide bounds on the gradients but construction of a flowsheet with Gradient intervals will provide the necessary information to allow for the Mean Value inclusion. This uses automatic differentiation with interval arithmetic to provide bounds on the gradients of the flowsheet variables with respect to the independent variables.

5.3. Optimisation of Extended Type Flowsheets

With the concept of extended types it is possible to see that Interval flowsheets are members of a class of possible models constructed with extended type Modules which use intervals to provide bounds on the outputs, modules based on automatic differentiation which provide gradients for the outputs and, using AD types based on intervals, provide bounds on the gradient of the outputs.

Clearly the optimisation algorithm is determined by the type, T, of the units in the flowsheet. An interval flowsheet can be optimised using an interval algorithm, a flowsheet based on convex underestimators can be optimised by one of the many algorithms based around convex underestimation [18,22,24].

Algorithms using convex underestimation are very effective for the solution of equation based models and should result in fewer bisections than an interval method. However, the solution of the underestimating NLP requires an evaluation of the flowsheet at each iteration. That is, the convex underestimators must be evaluated at every iteration of the NLP. This is not a difficulty in equation based models where the underestimating model is assumed to have been obtained by hand/symbolic manipulation. The model is fixed with the bounds, X^k, as parameters. In the modular case this advantage is lost. Each evaluation of the underestimating flowsheet *rebuilds* the convex underestimator model and then evaluates it.

This difficulty with convex underestimators arises because the modular flowsheet provides a method of evaluating the underestimators but does not actually provide the underestimators and because general convex underestimators cannot be easily characterised. A linear underestimator can easily be characterised by the coefficients and a linear model can be solved rapidly using Linear Programming.

So far two linear models have been proposed: the NE and the MV underestimators. The NE underestimators are linear under/over relaxations which can be used to obtain a linear relaxation of any variable in the flowsheet. Because they are linear they can easily be characterised by the coefficients of the different components and a linear programming problem can be constructed.

The MV relaxation also provides linear underestimators but the linear functions do not 'exist' inside the flowsheet. The underestimators are constructed from gradient information provided by the flowsheet in a manner similar to linearisation of the model. This method has the advantage that one set of gradient bounds obtained by AD can be used to construct as many underestimators as necessary, thereby reducing the total number of AD flowsheet evaluations. Each new underestimator requires one evaluation of the real arithmetic flowsheet.

Applying the approach taken with the equation based problems of constructing two underestimating planes at $\underline{x}$ and $\overline{x}$ requires one evaluation of the interval AD flowsheet which provides bounds on the variables in the flowsheet and bounds on the gradients with respect to the independent variables followed by one evaluation with real arithmetic for each of the points $\underline{x}$ and $\overline{x}$. This gives two underestimating planes and two overestimating planes.

In practice the overestimators are not as tight as the underestimators and this tends to hinder convergence because the relaxations of the product constraints are not tight. This suggests that it is better to make some extra evaluations with real arithmetic to obtain a tighter set of overestimators. Using the opposite corners of the box X^k provides a much better upper bound (overestimate) at the expense of extra real evaluations. This is especially advantageous when bilinear terms, x_1x_2, are involved because the underestimators constructed at $\underline{x_1}, \underline{x_2}$ and $\overline{x_1}, \overline{x_2}$ form the convex hull of x_1x_2 and the overestimators constructed at $\underline{x_1}, \overline{x_2}$ and $\overline{x_1}, \underline{x_2}$ form the concave hull of x_1x_2. Given that the Pooling problem is bilinear in the constraints this means that the extra evaluations required to construct MV underestimators at *each* corner will improve the performance overall because fewer interval AD evaluations need to be made. This has been found to be true and the extra planes improve the performance. These enhancements have been incorporated into the algorithms.

6. Examples

Two example problems will be presented. This first is the acyclic problem outlined above, the Haverly Pooling Problem. For this problem we have both the equation based model as well as the modular formulation. Results for both these are presented so that some idea of the premium paid for the convenience of the modular approach is shown. The second has a reactor with implicit equations and a recycle. Results are presented for the IGOR and RIGOR algorithms using the NE, MV and MVNE relaxations.

6.1. An Acyclic Problem

The Haverly Pooling problem given in Figure 1 is to blend four feeds into two products to minimise the total cost. This problem is nonconvex because of the bilinear terms involved in the splitter unit. The problem has an infinite number of local minimisers when the feed flowrates are zero, a strong local minimiser with an objective function value of -100, and a global minimiser at

$$F = [0, 100, 0, 0, 0, 100, 100, 200]^{\mathrm{T}}$$

and $x_A = 0.01$ with $f(x) = -400$.

The results of the equation based formulation of the pooling problem are given in Table 1. The results show that the MV and MVNE relaxations are superior to the NE relaxation and this is confirmed by other results [15]. The results of application to the modular pooling problem using IGOR and RIGOR with MV and MVNE relaxation are given in Table 2. The interval algorithm is applied with the Mean Value inclusion which utilises Interval Gradient types. The times are obtained on an Intel 486 with 8M of RAM using the Matlab implementation of (R)IGOR which will, necessarily, be slower than an implementation in C++. The RIGOR algorithm requires fewer iterations but in

Table 1. Equation based solution using different relaxations

	NE		MV		MVNE	
	Iter	Time (s)	Iter	Time (s)	Iter	Time (s)
IGOR	86	14.90	3	1.21	3	1.20
RIGOR	85	13.94	3	3.10	3	3.15

Table 2. Modular based solution using different relaxations

	MV		MVNE	
	Iter	Time (s)	Iter	Time (s)
IGOR	49	13.30	47	16.70
RIGOR	14	19.97	13	22.42

this case because of the extra overhead the computational time for solution is slightly longer.

Compared to the time taken to solve the Haverly Pooling problem in equation form the modular times are high. This is due to a number of factors:

- Evaluation of quantities through a modular flowsheet requires extra calculation as all intermediate results are calculated.
- Evaluation of the flowsheet results in calculation of some quantities, such as stream flowrates, which are not necessary in the EO form.
- In the EO form differentiation is performed symbolically and bounds are generated by interval arithmetic whereas the modular flowsheet uses bounding AD to calculate derivatives at each evaluation.

6.2. Recycle Flowsheet

This case study is taken from a synthesis problem in [28]. The operating parameters and sizes for one of the synthesis alternatives are optimised using the detailed models and costing information provided in [28].

Each unit has a capital cost, C_c and an operating cost, C_o which are incorporated into an objective function based on a pay-out time of 2.5 years.

The principle units are a CSTR and two separation columns. The models have been formulated in terms of component flowrates, $F_{s,i}$.

6.2.1. Reactor Model. The reactor is a continuous stirred tank reactor which models the reaction between chlorine and benzene (A) to produce monochlorobenzene (B) and Dichlorobenzene (C) at constant temperature. The reactor model has a single input stream, $F_{2,i}$, and three parameters, A_1, B_1 and V

$$\begin{aligned}
F_{3,A} &= F_{2,A} + V(-A_1), \\
F_{3,B} &= F_{2,B} + V(A_1 - B_1), \\
F_{3,C} &= F_{2,C} + V(A_1 - B_1)
\end{aligned}$$

the following constraints are used as residuals

$$\begin{aligned}
A_1 \sum_i F_{3,i} &= 0.413 F_{3,A}, \\
B_1 \sum_i F_{3,i} &= 0.055 F_{3,B}
\end{aligned}$$

and the capital cost is given by

$$C_c = 25.795 + 8.178V, \qquad C_o = 0.$$

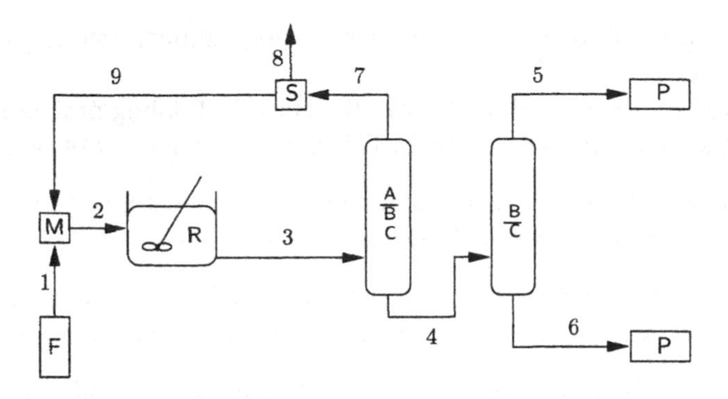

Figure 4. Problem with recycle.

Table 3. Modular solution of recycle problem

	Iter	Time (s)
IGOR	730	908.6
RIGOR	26	299.2

6.2.2. Columns. The distillation columns perform sharp splits between components. The costs are

C1

$$C_c = 132718 + 369F_{3,A} - 1114F_{3,B},$$
$$C_o = (21.67 + 4.65)39.1F_{3,A} + 10.7F_{3,B} + 3F_{3,C}.$$

C2

$$C_c = 25000 + 6985F_{4,B} - 1114F_{3,B},$$
$$C_o = (21.67 + 4.65)(55.662F_{4,B} + 26.212F_{4,C}).$$

Costs of the feeds and products are the purchase price of benzene, $27.98/kmol, and chlorine, $19.88/kmol, and the sale price of monochlorobenzene, $92.67/kmol, for which there is a demand of 50 kmol/hr.

Results for the application of the global optimisation algorithms to the different flowsheet Types are presented in Table 3. Again the RIGOR algorithm takes considerably fewer iterations but this time the computational time is also considerably better.

7. Conclusions

The results here on the acyclic problem show that there is a significant price to pay for the modular formulation. It has also been shown how reformulation and careful lower bounding strategies can make a big difference in the computational times required to solve the problems. It is anticipated that future developments in better bounding rules will continue to make interval based global optimisation an increasingly attractive approach for solving a wide range of problems. Our results also show that the RIGOR algorithm, with a LP reduction step for tightening the bounds, is a valuable enhancement over the IGOR algorithm.

Issues for general application within modular flowsheeting systems remain about how to deal with internal iterations within the unit models and how to convert existing code to handle intervals.

The modular approach to flowsheeting remains popular within the process industries. Although some issues remain, the approach presented here does make it possible to solve flowsheet optimisation using modular systems with interval based global optimisation algorithms.

Acknowledgements

The authors would like to acknowledge the financial support of the Engineering and Physical Sciences Research Council of the U.K. and the Centre for Process Systems Engineering.

Bibliography

[1] Edgar, T. F. and Himmelblau, D. M.: *Optimization of Chemical Processes*, Chemical Engineering Series, McGraw-Hill International, New York, 1988.

[2] Krotschek, C. and Narodoslawsky, M.: The sustainable process index. A new dimension in ecological evaluation, *Ecological Engineering* **6**(4) (1996), 241.

[3] Pistikopoulos, E. N., Stefanis, S. K. and Livingston, A. G.: A methodology for minimum environmental impact analysis, *American Institute of Chemical Engineers Symposium Series* **90**(303) (1994), 139–150.

[4] Steffens, M. A., Fraga, E. S. and Bogle. I. D. L.: Multicriteria process synthesis for generating sustainable and economic bioprocesses, *Comput. Chem. Engng.* **23** (1999), 1455–1467.

[5] Kallrath, J.: Mixed integer optimization in the chemical process industry: Experience, potential and future perspectives, *Trans. IChemE* **78**(Part A) (2000), 809–822.

[6] Biegler, L. T.: Simultaneous modular simulation and optimization, In: *Foundations of Computer-Aided Process Design, Proceedings of the Second International Conference held at Snowmass, Colorado*, 1983.

[7] Perkins, J. D.: Equation oriented flowsheeting, In: *Foundations of Computer Aided Process Design, Proceedings of the Second International Conference held at Snowmass, Colorado*, 1983.

[8] Grossmann, I. E. (ed.): *Global Optimization in Engineering Design*, Kluwer Acad. Publ., Dordrecht, 1996.

[9] Floudas, C. A.: Recent advances in global optimization for process synthesis, design and control: Enclosure of all solutions, *Computers Chem. Engng.* **23**(Suppl) (1999), S963–S973.

[10] Adjiman, C. S., Androulakis, I. P., Maranas, C. D. and Floudas, C. A.: A global optimisation method, αBB, for process design, *Computers Chem. Engng.* **20S** (1996), S419–S424.

[11] Vaidyanathan, R. and El-Halwagi, M.: Global optimisation of nonconvex programs via interval analysis, *Comput. Chem. Engng.* **18** (1994), 889.

[12] Smith, E. M. B. and Pantelides, C. C.: Global optimisation of general process models, In: I. E. Grossmann (ed.), *Global Optimisation for Engineering Design*, Kluwer Acad. Publ., Dordrecht, 1996, pp. 356–386.

[13] Kocis, G. R. and Grossmann, I. E.: Global optimization of nonconvex mixed-integer nonlinear-programming (MINLP) problems in process synthesis, *Ind. Eng. Chem. Res.* **27**(8) (1991), 1407–1421.

[14] Androulakis, I. P. and Venkatsubramanian, V.: A genetic algorithmic framework for process design and optimization, *Computers Chem. Engng.* **15**(4) (1991), 217–228.

[15] Byrne, R. P. and Bogle, I. D. L.: Global optimisation of constraiuned non-convex programs using reformulation and interval analysis, *Computers Chem. Engng.* **23** (1999), 1341–1350.

[16] Garrard, A. and Fraga, E. S.: Mass exchange network synthesis using genetic algorithms, *Computers Chem. Eng.* **22**(12) (1998), 1837–1850.

[17] Quesada, I. and Grossmann, I. E.: Global optimization algorithm for heat-exchanger networks, *Ind. Eng. Chem. Res.* **32**(3) (1993), 487–499.

[18] Quesada, I. and Grossmann, I. E.: Global optimization of bilinear process networks with multicomponent flows, *Comput. Chem. Engng.* **19**(12) (1995), 1219–1242.

[19] Moore, R. E.: *Interval Analysis*, Prentice-Hall, Englewood Cliffs, NJ, 1966.

[20] Hansen, E.: *Global Optimization Using Interval Analysis*, Marcel Dekker, New York, 1992.

[21] Ratschek, H. and Rockne, J.: *New Computer Methods for Global Optimization*, Ellis Horwood Ltd., Chichester, West Sussex, England, 1988.

[22] Ryoo, H. S. and Sahinidis, N. V.: Global optimization of nonconvex NLPs and MINLPs with applications in process design, *Computers Chem. Engng.* **19**(5) (1995), 551–566.

[23] Hansen, P., Jaumard, B. and Lu, S. H.: An analytical approach to global optimization, *Math. Programming* **52** (1991), 227–254.

[24] McCormick, G. P.: Computability of global solutions to factorable nonconvex programs: Part 1 – Convex underestimating problems, *Math. Programming* **10** (1976), 147–175.

[25] Rall, L. B.: *Automatic Differentiation: Techniques and Applications*, Lecture Notes in Comput. Sci., Springer-Verlag, Berlin, 1981.

[26] Byrne, R. P. and Bogle, I. D. L.: Global optimization of modular process flowsheets, *Ind. Eng. Chem. Res.* **39** (2000), 4296–4301.

[27] Byrne, R. P. and Bogle, I. D. L.: Global optimisation using interval analysis, In: I. E. Grossmann (ed.), *Global Optimisation for Engineering Design*, Kluwer Acad. Publ., Dordrecht, 1996, pp. 155–174.

[28] Floudas, C. A.: *Nonlinear and Mixed-Integer Optimisation. Fundamentals and Applications*, Oxford University Press, New York, 1995.

[29] Vaidyanathan, R. and El-Halwagi, M.: Global optimisation of MINLPs by interval analysis, In: I. E. Grossmann (ed.), *Global Optimisation for Engineering Design*, Kluwer Acad. Publ., Dordrecht, 1996, pp. 175–195.

Chapter 4

ONE-DIMENSIONAL GLOBAL OPTIMIZATION BASED ON STATISTICAL MODELS

James M. Calvin
Department of Computer and Information Science
New Jersey Institute of Technology
Newark, NJ 07102-1982, USA

Antanas Žilinskas
Institute of Mathematics and Informatics, VMU
Akademijos str. 4
Vilnius, LT2600, Lithuania

Abstract This paper presents a review of global optimization methods based on statistical models of multimodal functions. The theoretical and methodological aspects are emphasized.

Keywords: Optimization, statistical models, convergence

1. Introduction

Global optimization is one of the most important subjects for applications and at the same time one of the most difficult subjects in optimization and even computing in general. Because of the practical importance of the field and the shortage of mathematical theory (compared with local optimization theory), many heuristic methods were proposed by the experts of different applied subjects.

Mathematical investigation in the field began using a standard approach of computing: a class of problems is defined by means of postulating features of an objective function and convergence of a method is investigated. A class of Lipshitz functions with known Lipshitz constant is the most popular class of functions. Special methods were investigated for minimization of concave and

G. Dzemyda et al. (eds.), Stochastic and Global Optimization, 49–63.
© 2002 *Kluwer Academic Publishers.*

quadratic functions, for example. Although in many papers it is not formally emphasized, the minimax approach was used. In light of pessimistic results implying exponential complexity of the proposed algorithms, investigation of optimal algorithms became important. However, it was proved that optimal algorithms also have exponential complexity. The minimax approach can not escape exponential complexity because a broad class of multimodal functions contains pathological examples. Moreover, it became clear that in the minimax setting adaptation can not help.

The first paper on a global optimization algorithm justified by arguments of average rationality was [1]. However, extension of the model used by H. Kushner to the multi-dimensional case was difficult. The model also appeared to be inappropriate for many real optimization problems because of the non-differentiability of the sample functions. The extension of the method to smooth function models was also prevented by computational difficulties. Therefore, the development of a global optimization approach based on statistical models from the very beginning seemed difficult theoretically as well as from the point of view of algorithmic implementation.

The problem of optimal algorithms became important for the statistical model approach as well as for the minimax approach. Bayesian methods were formulated by J. Mockus in [2], and further developed in [3].

In this paper we present a review of one-dimensional global optimization methods based on statistical models. In the one-dimensional case the implementation is much closer to the theoretical schema than in the multi-dimensional case. The reviews of earlier results may be found in the following publications covering broader subjects on statistical models in global optimization: [3–8].

2. Statistical Models

The choice of a statistical model for constructing global optimization algorithms should be justified by the usual arguments of adequacy and simplicity. The theory of stochastic processes is well developed from the point of view of analysis. However, it is not so rich from the point of view of synthesis. For example, the properties of a probability measure concentrated on the class of unimodal functions have been not yet investigated. Therefore, a well understood stochastic function, whose properties are consistent with the general information on the target problem, is normally chosen for the statistical model if it is acceptably simple for algorithmic implementation.

The Wiener process is a popular model for constructing one-dimensional global optimization algorithms; it was used in [1,9–16]. Besides global optimization, the Wiener model was also used in investigations of the average complexity of integration and interpolation problems [17]. If a target problem is expected to be very complicated, with many local minima, then the assumption of independence of the increments of function values for sufficiently remote disjoint subintervals seems acceptable. This is the main argument in favor of the global adequacy of the Wiener process to model complicated multimodal problems. The Wiener process is favorable also from the computational point of view because it is Markovian. However, the sampling functions of the Wiener process are not differentiable with probability one, almost everywhere. This feature draws criticism for the use of the Wiener process for a model because objective functions with such severe local oscillations do not arise in many applications. Summarizing the advantages and disadvantages, the Wiener process seems justified for a global description of an objective function but not as a local model. The latter conclusion has motivated the introduction of a dual statistical/local model where the global behavior of an objective function is described by the Wiener process, and its behavior in small subintervals of the main local minimizers is described by a quadratic function [10]. Several versions of the algorithms based on the dual model are described in [10,12,14].

Besides the properties of the Wiener process discussed above, its advantage is availability of an analytical formula for the probability distribution of the minimum and its location [18,19]. Using the latter, a stopping condition may be constructively defined such that the probability of finding the global minimum exceeds a predefined level. Such a stopping condition is implemented in the algorithms based on the dual statistical/local models discussed above.

Because the Wiener process has the mentioned disadvantages as a model of objective functions but also has serious computational advantages, it seems reasonable to derive Wiener process based models. The sample functions of the integrated Wiener process

$$\eta(t) = \int_0^t \xi(\tau)\, d\tau,$$

where $\xi(\tau)$ is the Wiener process, are differentiable with probability one. This process is Markovian with respect to the observation of vectors composed from the function value and its derivative. Such a model was used for construction of global optimization algorithms in [20]. For local optimization a very important characteristic of an objective function is the second derivative. To get a statistical model with twice differentiable sample functions the Wiener process may be twice integrated.

On the other hand, a stationary stochastic process may be chosen for a model, guaranteeing the desired smoothness of sample functions by an appropriate choice of correlation function. It is well known that the smoothness of the sample functions of a stationary stochastic process is determined by the behavior of the correlation function in a neighborhood of zero. The correlation function of the unique Markov stationary stochastic process is $\exp(-c\,|t|)$, where $c > 0$, implying nondifferentiability of sample functions. The smooth stationary stochastic processes were outside of the attention of researchers in global optimization because of implementation difficulties in the non-Markovian case. However, an approximated version of the P-algorithm for a smooth function model, proposed in [21], is of the same complexity as the P-algorithm for the Wiener model. This result may change the unfavorable attitude towards stationary stochastic processes as the statistical models for global optimization.

The stochastic models considered above are well known. However, the properties investigated in the general theory of stochastic processes may be not very helpful for the construction of global optimization algorithms, and conversely, important properties for the construction of global optimization algorithms are largely outside the interest of probability theoreticians. A stochastic process is proposed in [22] as a specific model for an information-based approach to global optimization. The stochastic process ξ_i is defined by means of independent random increments, where the feasible interval is discretized by the points $i/N, i = 0, \ldots, N$, and the increments $\xi_i - \xi_{i-1}$, are supposed independent Gaussian random variables with variance σ^2, and average $-m$ for $i < \alpha$, and m for $i \geqslant \alpha$. The parameter α is supposed to be a random variable with known (normally uniform) distribution on $\{0, 1/N, 2/N, \ldots, 1\}$. The aposteriori distribution of α may be calculated with respect to observed function values. The most likely value of α may be close to the minimizer of the sample function, e.g., this is true for $m \gg \sigma$. But in the latter case the adequacy of the model to global optimization problems is doubtful since the sample functions of the stochastic process most likely are unimodal. Anyway, the maximization of information on the location of α in the frame of the considered model may be an efficient procedure to search for the global minimum.

The available information on the target objective functions is not always sufficient to define a probability measure on the set of functions. In some papers on local minimization and search for a zero [23–26] the statistical assumptions are combined with deterministic assumptions without the definition of a probability measure. Similar combination of deterministic and statistical assumptions is used to construct a one-dimensional global minimization algorithm in [27]. The assumptions on objective function values related to dispersion analysis justify a method of structure analysis in [28].

The introduction of general statistical models, including stochastic processes as well as statistical models without definition of a probability measure on the set of potential objective functions, opens the possibility of a unified approach to global optimization based on the ideas of rational decision theory. The uncertainty regarding values of an objective function normally satisfy the assumptions of rationality, implying acceptability of a random variable as a model of the unknown function value. Correspondingly, the family of random variables $\xi_x, 0 \leqslant x \leqslant 1$, is acceptable as a statistical model of a class of objective functions [29–31]. The further normative modeling [32] of uncertainty on behavior of an objective function concerns extrapolation under uncertainty and results in definition of characteristics of a family of random variables [33]. Such a class of models includes computationally simpler models than stochastic processes. It is interesting to note that by means of adding a few informal assumptions to the general postulates on uncertainty, a statistical model corresponding, e.g., to the Wiener process, may be constructed [5].

Statistical models may be equivalent to deterministic models in the sense of algorithms constructed using these models. For example, a P-algorithm based on the Wiener process model coincides with the one-dimensional algorithm based on a radial function model [34].

The theoretical justification and computational details of the statistical models used in one-dimensional global optimization are presented in [1,27,29,35–37].

3. Methods

Let a stochastic process $\xi(x)$, $0 \leqslant x \leqslant 1$, be chosen as a model of the target objective functions. With such a model it is implicitly accepted as the intention to minimize many functions with similar characteristics. Therefore, the average-case optimal algorithm with respect to the stochastic process is interesting. The notion of average optimality is related to the average error of the algorithm. Let us denote a class of minimization algorithms by $\mathcal{A}$, and $\xi_{A,n}$ the estimate of the global minimum obtained after n iterations of the algorithm $A \in \mathcal{A}$ while minimizing a randomly chosen sample function $\xi(x)$. Very general assumptions guarantee existence of the mean error of the estimate of the global minimum $\varepsilon(A, n)$

$$\varepsilon(A, n) = \mathbf{E}\{\xi_{A,n} - \xi_0\},$$

where $\mathbf{E}$ denotes the expectation operator and ξ_0 denotes the minimum value of the sample function. Let the algorithms in $\mathcal{A}$ define the current observation point taking into account all function values and observation points at previous minimization steps. For the case of an *apriori* fixed number N of observations the optimal (Bayesian) algorithm

$$\varepsilon(B, N) = \min_{A \in \mathcal{A}} \varepsilon(A, N), \tag{3.1}$$

is defined by the solution of a system of recurrent equations, as shown in [2]. However, the complexity of the latter system prevents analytical or even numerical investigation of the optimal algorithms.

Because of implementation difficulties of optimal algorithms semi-optimal algorithms were investigated. A one-step Bayesian algorithm was proposed in

[2] as a simplified version of the Bayesian algorithm. The results of investigation of the Bayesian approach in global optimization are presented in [3].

A version of a one-dimensional one-step Bayesian algorithm based on the Wiener model was implemented and investigated in [38]. It was shown that the general definition of one-step Bayesian algorithm

$$x_{k+1} = \arg\min_{0 \leqslant x \leqslant 1} \mathbf{E}\{\min(\xi(x), \xi_{0k}) | \xi(x_1), \ldots, \xi(x_k)\},$$
$$\xi(x_{0k}) = \min\{\xi(x_1), \ldots, \xi(x_k)\},$$

may be reduced to the solution of $k-1$ unimodal problems. Experimental investigation has shown that the one-step algorithm performs a very local search concentrating the observations in close vicinities of the best found points [38]. This property may be well explained as a consequence of the one-step optimality: if only one observation is planned then it seems not rational to risk but better to make an observation near to the best point, where an improvement (maybe only small) is likely. However, such an explanation does not help very much to justify modifications of the one-step Bayesian algorithm improving globality of search. In [38] the efficiency of search is improved by means of modification of the parameters of the statistical model, but it is difficult to justify such a modification theoretically.

A family of random variables is a more general statistical model than a stochastic process. The former frequently may be accepted in the case when the latter can not be constructed because of insufficiency or inconsistency of the available information. The definition of a rational optimization method with respect to such a general model is not so obvious as (3.1). The axiomatic approach is used to define the rational decisions in complicated situations of decision making. The choice of a point to evaluate the objective function at the current minimization step corresponds to a situation of complicated decision making. In [35] such a choice is justified by the axioms of rationality, and it is proved that for compatibility with the rationality axioms the choice of the current point should be

$$x_{k+1} = \arg\max_{0 \leqslant x \leqslant 1} \mathbf{P}\{\xi(x) \leqslant (\xi_{0k} - \varepsilon_k) | \xi(x_1), \ldots, \xi(x_k)\}. \tag{3.2}$$

It is obvious that the algorithm (3.2) may be applied also in the case of a stochastic process used for a model of minimization problems. An *ad hoc* algorithm (3.2) with the Wiener model and observations in the presence of noise was proposed in [1]. The character of globality/locality of (3.2) may be regulated by means of the choice of an aspiration level ε_k: for large ε_k the search is more global than for small ε_k [35]. Different one-dimensional versions of (3.2) based on different statistical models have been considered in [1,5,8–11,14,20,21,38–41].

In [36] it is proposed to alternate a few steps of (3.2) with a few steps of choosing the points according to

$$x_{k+1} = \arg \min_{0 \leqslant x \leqslant 1} \mathbf{E}\{\xi(x)|\xi(x_1), \ldots, \xi(x_k)\}.$$

Minimization of the conditional mean is called the localization phase. However, the latter phase may degenerate, as is discussed in [5]; i.e., the minimizer of the conditional mean may coincide with the best found point. The introduction of the localization phase seems not well justified also because the locality/globality of search may be easily regulated by means of changing the aspiration level as it was mentioned above.

In the information approach a special stochastic process is constructed whose parameter α is considered as a carrier of information on the global minimum point. The information algorithm originally was constructed in [22], where the current point was chosen to maximize information on α. Some modifications of this algorithm are proposed in [6] where the criterion of information is substituted with the criterion of average loss, e.g., with a step-wise loss function. The algorithm that is optimal with respect to such a loss performs the current observation of the objective function at the maximally likely point of α

$$x_{k+1} = j/N, \quad j = \arg \max_{0 \leqslant i \leqslant N} p_k(i),$$
$$p_k(i) = \mathbf{P}\{\alpha = i/N|\xi(x_1), \ldots, \xi(x_k)\}.$$

Frequently the solution obtained by means of a global optimization algorithm is calculated more precisely by means of a local optimization algorithm. Usually such a combination is heuristic. The statistical model of the objective function enables one to justify the combination of global and local algorithms in the frame of statistical methodology. For example, in parallel with global search based on the Wiener model a statistical hypothesis is tested on finding a vicinity of a local minimum [10,12,14]. If such a vicinity is found, a current local minimizer is calculated with predefined accuracy by means of a local minimization algorithm. The detected vicinity is excluded from further global search, and local and global strategies complement their strengths.

Different aspects of the algorithmic implementation of global optimization methods based on statistical models are considered in [1,5,6,9–12,14,15,36, 42–44].

An interesting and important generalization of the methods discussed above would be for the case of function values corrupted by noise. The inclusion of noise into the statistical model does not cause fundamental difficulties since the information on function values is integrated using conditional probability distributions. However, computational difficulties occur because of loss of the Markovian property. An algorithm based on the Wiener model and its codes are presented in [13,14].

4. Convergence

The algorithms considered above have been constructed as optimal procedures with respect to different statistical models. The algorithms may be used for a very broad class of problems, and the adequacy of the models to the real data is normally not tested. Although not optimal, the algorithms may nevertheless be competitive. Therefore, their performance is interesting under weaker assumptions than used to define a statistical model.

In global optimization problems the objective function rather frequently is given by means of a code implementing a complicated algorithm. While considering the convergence conditions the assumptions on the function should be rather weak. It seems that continuity of the objective function is the weakest reasonable assumption. To guarantee convergence for any continuous function the algorithm should generate a dense sequence of trial points. The same necessary and sufficient convergence condition is valid also for more narrow classes of objective functions, e.g., the class of Lipshitz functions with *a priori* unknown Lipshitz constant; for a discussion see [45,46].

Some subintervals of the feasible interval may be excluded from consideration only in the case that guaranteed lower bounds on function values are available, e.g., based on a known Lipshitz constant for function values or derivatives, or interval arithmetic based computations. Sometimes the convergence theorems do not explicitly assume that the Lipshitz constant is known, but this assumption is implicit. In [47] it is assumed (Theorem 3, iii) that the parameter of the algorithm m is larger than L, the actual Lipshitz constant of the derivative of the objective function. Similar assumptions are frequently made also in the case when adaptive estimates of a Lipshitz constant are used. For example, in [48] the assumptions are made (Theorem 2) that the local tuned Hölder constants should satisfy the inequalities involving unknown global minimum and minimizer. To have guaranteed convergence for any continuous function not depending on the choice of parameters and/or run time conditions, everywhere dense sequences of trial points should be required.

To prove the sequence (3.2) everywhere dense it is necessary to prove that every subinterval will be eventually chosen and subdivided by the new trial point with bounded ratio. Normally these properties may be easily proved as consequences of the properties of conditional mean and conditional variance of the statistical model used to justify the considered algorithm [21,35,38,39,41]. A comparative study of convergence of different one-dimensional algorithms, including some statistical model based algorithms, may be found in [49].

For the convergence proofs we refer to [21,35,38,39,41,44,47,50,51]. Since convergence does not imply efficiency, the practical efficiency of the algorithm is tested by computational experiments. As a theoretical criterion of efficiency normally the convergence rate is considered.

5. Convergence Rates

In this section we consider the rates at which the approximation error converges to zero for various algorithms.

Our goal is to approximate the global minimum $f^* = \min_{0 \leqslant s \leqslant 1} f(s)$ of the objective function f which is a member of a class F of at least continuous functions; i.e., we assume $F \subset C([0, 1])$. Let $t_1, t_2, \ldots$ be the observation points, and let $\Delta_n = \min_{1 \leqslant i \leqslant n} f(t_i) - f^*$ be the error after n function evaluations. If an algorithm is based on a probability model P on F and each t_{n+1} is chosen as a function of only the previous n function evaluations, then $\{\Delta_n : n \geqslant 0\}$ is a sequence of random variables, and it is natural to examine the rate at which they decrease in a probabilistic sense. This is not the only aspect we are interested in; we also consider the convergence rate for arbitrary elements of F. We take the view that it may be appropriate to base an algorithm on a statistical model even if it is intended to apply the algorithm to objective functions that are not drawn according to the probability distribution.

Ritter [17] established the best possible convergence rate of any nonadaptive optimization algorithm under the Wiener measure. If each point t_n is chosen independently of the function evaluations, then the error is $\Omega(n^{-1/2})$. The convergence rate of $\Theta(n^{-1/2})$ is attained for equispaced points. That is, if the number of evaluations n is fixed in advance, then $t_i = i/n$ is order-optimal.

There are many nonadaptive algorithms with errors $\Delta_n = \Theta(n^{-1/2})$. Let us begin with randomized algorithms. The minimizer of the Wiener process is almost surely unique and has the arcsine density

$$\zeta(t) = \frac{1}{\pi\sqrt{t(1-t)}}$$

on $(0, 1)$. Thus a natural choice for a randomized algorithm is to choose the t_i independently according to the distribution ζ. With this choice,

$$\sqrt{n}E(\Delta_n) \to \frac{1}{\sqrt{2\pi}}\mathcal{B}(3/4, 3/4) \approx 0.675978,$$

where $\mathcal{B}$ is the beta function. This is not the best choice of sampling distribution; if the t_i are instead chosen independently according to the $\mathcal{B}(2/3, 2/3)$ distribution, then

$$\sqrt{n}E(\Delta_n) \to \frac{1}{\pi\sqrt{2}}\mathcal{B}(2/3, 2/3)^{3/2} \approx 0.662281,$$

a slight improvement. Both distributions are better than the uniform distribution, for which

$$\sqrt{n}E(\Delta_n) \to \frac{1}{\sqrt{2}} \approx 0.707107.$$

These results are discussed in [52].

The deterministic versions of the above nonadaptive algorithms have slightly better convergence rates. A sequence of knots $\{t_i^n; 1 \leqslant i \leqslant n\}$ is a *regular sequence* if the knots form quantiles with respect to a given density ψ; i.e.,

$$\int_{t=0}^{t_i^n} \psi(t)\,\mathrm{d}t = \frac{i-1}{n-1}$$

for $1 \leqslant i \leqslant n$; see [53]. If we take ψ to be the uniform distribution and construct an algorithm based on the corresponding regular sequence, then

$$\sqrt{n}E(\Delta_n) \to c \approx 0.5826.$$

With ψ the arcsine density, the convergence is to a value of approximately $0.956c$, and with ψ the $\mathcal{B}(2/3, 2/3)$ density, this improves to approximately $0.937c$; see [54]. The deterministic algorithms described above are noncomposite in that the number of evaluations n must be specified in advance; this can be seen as a disadvantage relative to the randomized algorithms that tends to offset their better convergence rates; see [52].

Some of these results on nonadaptive algorithms have been extended to diffusion processes other than Wiener process [55]. These results include showing that the normalized error converges in distribution (for both random and deterministic nonadaptive algorithms) when the path is random, but pathwise, the normalized error fails to converge in distribution for almost all paths [16].

In [40], the P-algorithm based on the scheme with constant $\varepsilon_k = \varepsilon > 0$ is studied under the Wiener measure. A sequence of stopping times $\{n_k\}$ is constructed so that

$$\lim_{k\to\infty} P\left(\frac{\sqrt{n_k}}{\varepsilon\sqrt{\rho}}\Delta_{n_k} \leqslant y\right) = G(y),$$

as $k \to \infty$, where

$$\rho = \int_{t=0}^{1} \frac{\mathrm{d}t}{(f(t) - f^* + \varepsilon)^2}$$

and where G is the distribution function of the minimum of a two-sided three-dimensional Bessel process over a lattice of diameter 1 with $U(0, 1)$ offset.

Most algorithms in use are adaptive in that each function evaluation depends on the results of previous evaluations. In the worst case setting, adaptation does not help much under quite general assumptions. If F is convex and symmetric (in the sense that $-F = F$), then the maximum error under an adaptive algorithm with n observations is not smaller than the maximum error of a non-adaptive method with $n + 1$ observations; see [56]. Thus adaptive methods in common use can not be justified by a worst case analysis, which motivates our interest in the average-case setting.

By allowing the $\{\varepsilon_n\}$ to depend on the past observations instead of being a fixed deterministic sequence, it is possible to establish a much better convergence rate than that of the algorithm described above. In [57] an algorithm was constructed with the property that the error converges to zero for any continuous function and furthermore, the error is of order e^{-nc_n}, where $\{c_n\}$ (a parameter of the algorithm) is a deterministic sequence that can be chosen to approach zero at an arbitrarily slow rate. Notice that the convergence rate is now almost exponential in the number of observations n. The computational cost of the algorithm grows quadratically, and the storage increases linearly, since all past observations must be stored.

In [21] an approximate version of the P-algorithm was proposed for a model on $F = C^2([0, 1])$. A fixed threshold $\varepsilon > 0$ was chosen, and the convergence rate $\Delta_n = O(n^{-2})$ was established for any element of F with positive second derivative at the global minimizer. In [41] this algorithm was improved to use a decreasing sequence of thresholds $\varepsilon_n \downarrow 0$. The sequence $\varepsilon_n = n^{-1+\delta}$, for a small positive δ, results in a convergence rate of $\Delta_n = O(n^{-3+\delta})$.

In [20] versions of the P-algorithm are investigated for the integrated Wiener measure on $F = C^2([0, 1])$. Versions were considered for function evaluations only and also for function and derivative evaluations.

In [58] a class of adaptive algorithms was introduced that operate with memory of cardinality 2. Within this class, for any $\delta > 0$, it is shown that an algorithm can be constructed that converges to the global minimum at rate $n^{-1+\delta}$ in the number of observations n. More precisely,

$$P\big(n^{1-\delta/2}\Delta_n \leqslant x\big) \to \tanh^2(\sqrt{2}x), \quad x > 0.$$

This rate is in contrast to the $n^{-1/2}$ rate that is the best achievable with a non-adaptive algorithm. A useful property of this class of algorithm is that they can be implemented in parallel on two processors with minimal communication. The limiting joint distribution of the normalized function and location error variables is derived. The construction of the algorithms does not depend on special properties of the Wiener measure and the algorithm class generalizes to memory of arbitrary (finite) cardinality m. It is possible to choose m and an algorithm within the class such that, under Wiener measure, the error decreases to zero faster than any fixed polynomial in n^{-1}.

6. Testing and Applications

One of the aims of experimental testing of optimization algorithms is to demonstrate the theoretically predicted behavior of the algorithm, and to understand its properties which are difficult to analyze theoretically. This kind of testing is interesting mainly to the researchers in optimization. The users are mainly interested in comparative testing whose results might justify the choice of the most appropriate algorithm. Two criteria are most important in the comparative testing: reliability and efficiency. The testing results in the papers of different authors are difficult to compare. There are objective difficulties implied by different assessment of two criteria by different authors. One of the factors influencing the trade-off is termination conditions, which are different in different algorithms. Some additional difficulty is caused by different test functions used. Generally speaking, the algorithms based on statistical models compete favorably with the algorithms based on the other approaches [5,15]. Recently the testing methodologies and sets of test functions are discussed more extensively [5,49,59,60], and this may speed up the standardization of testing methodology.

Practical problems are seldom one-dimensional. An example of a one-dimensional problem of statistical estimation solved by means of the Wiener model based algorithm [38] is presented in [61]. Sometimes the one-dimensional algorithm is used cycling the variables. Such a coordinate-wise extension of a one-dimensional algorithm to multi-dimensional problems may be efficient only in the case of weak dependence of the variables. Examples of successful application of the coordinate method to practical multimodal problems are presented in [5,28,62]. Let us note that coordinate-wise global optimization is easily parallelized [42].

There are examples of successful practical applications of one-dimensional algorithms extended to multi-dimensional problems by means of dynamic programming and mapping of multi-dimensional sets to one-dimensional intervals [6]. The first type of extension is defined by the following equality

$$\min_{X \in A} f(X) = \min_{x_n^- \leqslant x_n \leqslant x_n^+} \Big(\dots \Big(\min_{x_1^- \leqslant x_1 \leqslant x_1^+} f(X) \Big) \dots \Big), \tag{6.1}$$

where $A = X : x_i^- \leqslant x_i \leqslant x_i^+, i = 1, \dots, n$. The disadvantage of this method is explicitly exponential growth of number of function calls with increase of n. More efficient extension of one-dimensional methods to multi-dimensional case is based on Peano type space filling curves [6,44,49]. Maybe, most powerful extension of one-dimensional methods is indirect, i.e. in [50,63,64] multi-dimensional methods are constructed by means of generalization of properties of one-dimensional algorithms to the multi-dimensional case.

Bibliography

[1] Kushner, H.: A versatile stochastic model of a function of unknown and time-varying form, *J. Math. Anal. Appl.* **5** (1962), 150–167.

[2] Mockus, J.: On Bayesian methods of search for extremum, *Automatics and Computers* **3** (1972), 53–62 (in Russian).

[3] Mockus, J.: *Bayesian Approach to Global Optimization*, Kluwer Acad. Publ., Dordrecht, 1989.

[4] Boender, G. and Romeijn, E.: Stochastic methods, In: R. Horst and P. Pardalos (eds), *Handbook of Global Optimization*, Kluwer Acad. Publ., Dordrecht, 1995, pp. 829–869.

[5] Törn, A. and Žilinskas, A.: *Global Optimization*, Springer, 1989.

[6] Strongin, R.: *Numerical Methods in Multiextremal Optimization* (in Russian), Nauka, 1978.

[7] Horst, R. and Tuy, H.: *Global Optimization – Deterministic Approaches*, 2nd edn, Springer, 1992.

[8] Zhigljavski, A. and Žilinskas, A.: *Methods of Search for Global Extremum* (in Russian), Nauka, Moscow, 1991.

[9] Kushner, H.: A new method of locating the maximum point of an arbitrary multipeak curve in the presence of noise, *J. Basic Eng.* **86** (1964), 97–106.

[10] Žilinskas, A.: On global one-dimensional optimization, *Engrg. Cybernet., Izv. AN USSR* **4** (1976), 71–74 (in Russian).

[11] Archetti, F. and Betro, B.: A probabilistic algorithm for global optimization, *Calcolo* **16** (1979), 335–343.

[12] Žilinskas, A.: Optimization of one-dimensional multimodal functions, algorithm 133, *Appl. Statist.* **23** (1978), 367–385.

[13] Žilinskas, A.: Mimun-optimization of one-dimensional multimodal functions in the presence of noise, algoritmus 44, *Apl. Mat.* **25** (1980), 392–402.

[14] Žilinskas, A.: Two algorithms for one-dimensional multimodal minimization, *Mathematische Operationsforschung und Statistik, ser. Optimization* **12** (1981), 53–63.

[15] Locatelli, M.: Baeysian algorithms for one-dimensional global optimization, *J. Global Optim.* **10** (1997), 57–76.

[16] Calvin, J. M. and Glynn, P. W.: Average case behavior of random search for the maximum, *J. Appl. Probab.* **34** (1997), 631–642.

[17] Ritter, K.: Approximation and optimization on the Wiener space, *J. Complexity* **6** (1990), 337–364.

[18] Shepp, L. A.: The joint density of the maximum and its location for a Wiener process with drift, *J. Appl. Probab.* **16** (1976), 423–427.

[19] Lindgren, G.: Local maxima of Gaussian fields, *Ark. Math.* **10** (1972), 195–218.

[20] Calvin, J. M. and Žilinskas, A.: On convergence of a p-algorithm based on a statistical model of continuously differentiable functions, *J. Global Optim.* **19** (2001), 229–245.

[21] Calvin, J. M. and Žilinskas, A.: On convergence of the p-algorithm for one-dimensional global optimization of smooth functions, *J. Optim. Theory Appl.* **102**(3) (1999), 479–495.

[22] Neimark, J. and Strongin, R.: Information approach to search for minimum of a function, *Izv. AN SSSR, Engineering Cybernetics* No. 1 (1966), 17–26 (in Russian).

[23] Fine, T.: Optimal search for location of the maximum of a unimodal function, *IEEE Trans. Inform. Theory* (2) (1966), 103–111.

[24] Converse, A.: The use of uncertainty in a simultaneos search, *Oper. Res.* **10** (1967), 1088–1095.

[25] Heyman, M.: Optimal simultaneos search for the maximum by the principle of statistical information, *Oper. Res.* (1968), 1194–1205.

[26] Neuman, P.: An asymptotically optimal procedure for searching a zero or an extremum of a function, In: *Proceedings of 2nd Prague Symp. Asymp. Statist.*, 1981, pp. 291–302.

[27] Timonov, L.: An algorithm for search of a global extremum, *Engrg. Cybernet.* **15** (1977), 38–44.

[28] Shaltenis, V.: *Structure Analysis of Optimization Problems* (in Russian), Mokslas, Vilnius, 1989.

[29] Žilinskas, A.: On statistical models for multimodal optimization, *Mathematische Operationsforschung und Statistik, ser. Statistics* **9** (1978), 255–266.

[30] Žilinskas, A.: Axiomatic approach to statistical models and their use in multimodal optimization theory, *Math. Programming* **22** (1982), 104–116.

[31] Žilinskas, A.: A review of statistical models for global optimization, *J. Global Optim.* (1992), 145–153.

[32] Luce, D. and Suppes, P.: Preference, utility and subjective probability, In: R. Bush, D. Luce and E. Galanter (eds), *Handbook of Global Optimization*, Wiley, 1965, pp. 249–410.

[33] Žilinskas, A.: Axiomatic approach to extrapolation problem under uncertainty, *Automatics and Remote Control*, No. 12 (1979), 66–70 (in Russian).

[34] Gutman, H.-M.: A radial basis function method for global optimization, *J. Global Optim.* **19** (2001), 201–227.

[35] Žilinskas, A.: Axiomatic characterization of a global optimization algorithm and investigation of its search strategies, *Oper. Res. Lett.* **4** (1985), 35–39.

[36] Shagen, I.: Internal modelling of objective functions for global optimization, *J. Optim. Theory Appl.* **51** (1986), 345–353.

[37] Žilinskas, A.: Statistical models for global optimization by means of select and clone, *Optimization* **48** (2000), 117–135.

[38] Žilinskas, A.: One-step Bayesian method for the search of the optimium of one-variable functions, *Cybernetics*, No. 1 (1975), 139–144 (in Russian).

[39] Calvin, J. M. and Žilinskas, A.: On the choice of statistical model for one-dimensional p-algorithm, *Control and Cybernet.* **29**(2) (2000), 555–565.

[40] Calvin, J. M.: Convergence rate of the p-algorithm for optimization of continuous functions, In: P. Pardalos (ed.), *Approximation and Complexity in Numerical Optimization: Continuous and Discrete Problems*, Kluwer Acad. Publ., Boston, 1999, pp. 116–129.

[41] Calvin, J. M. and Žilinskas, A.: A one-dimensional p-algorithm with convergence rate $o(n^{-3+\delta})$ for smooth functions, *J. Optim. Theory Appl.* **106** (2000), 297–307.

[42] Törn, A. and Žilinskas, A.: Parallel global optimization algorithms in optimal design, *Lecture Notes in Control and Inform. Sci.* 143, Springer, 1990, pp. 951–960.

[43] Locatelli, M. and Schoen, F.: An adaptive stochastic global optimization algorithm for one-dimensional functions, *Ann. Oper. Res.* **58** (1995), 263–278.

[44] Sergeyev, Y.: An information global optimization algorithm with local tuning, *SIAM J. Optim.* **5** (1995), 858–870.

[45] Žilinskas, A.: Note on Pinter's paper, *Optimization* **19** (1988), 195.

[46] Žilinskas, A.: A note on extended univariate algorithms by J. Pinter, *Computing* **41** (1989), 275–276.

[47] Gergel, V. and Sergeev, Y.: Sequential and parallel algorithms for global minimizing functions with Lipshitz derivatives, *Internat. J. Comput. Math. Appl.* **37** (1999), 163–179.

[48] Sergeyev, Y.: Parallel information algorithm with local tuning for solving multidimensional global optimization problems, *J. Global Optim.* **15** (1999), 157–167.

[49] Hansen, P. and Jaumard, B.: Lipshitz optimization, In: R. Horst and P. Pardalos (eds), *Handbook of Global Optimization*, Kluwer Acad. Publ., Dordrecht, 1995, pp. 404–493.

[50] Pinter, J.: Extended univariate algorithms for n-dimensional global optimization, *Computing* **36** (1986), 91–103.

[51] Hansen, P. and Jaumard, B.: On Timonov's algorithm for global optimization of univariate Lipshitz functions, *J. Global Optim.* **1** (1991), 37–46.

[52] Al-Mharmah, H. and Calvin, J. M.: Optimal random non-adaptive algorithm for global optimization of brownian motion, *J. Global Optim.* **8** (1996), 81–90.

[53] Ritter, K.: *Average-Case Analysis of Numerical Problems*, Lecture Notes in Math. 1733, 2000.

[54] Calvin, J. M.: Average performance of passive algorithms for global optimization, *J. Math. Anal. Appl.* **191** (1995), 608–617.

[55] Calvin, J. M. and Glynn, P. M.: Complexity of non-adaptive optimization algorithms for a class of diffusions, *Comm. Statist. Stochastic Models* **12** (1996), 343–365.

[56] Novak, E.: *Deterministic and Stochastic Error Bounds in Numerical Analysis*, Lecture Notes in Math. 1349, Springer, 1988.

[57] Calvin, J. M.: A one-dimensional optimization algorithm and its convergence rate under the wiener measure, *J. Complexity* (2001), accepted.

[58] Calvin, J. M.: Average performance of a class of adaptive algorithms for global optimization, *Ann. Appl. Probab.* **7** (1997), 711–730.

[59] Törn, A., Viitanen, S. and Ali, M.: Stochastic global optimization: Problem classes and solution techniques, *J. Global Optim.* **14** (1999), 437–447.

[60] Floudas, C. and Pardalos, P.: *Handbook of Test Problems in Local and Global Optimization*, Kluwer Acad. Publ., Dordrecht, 1999.

[61] Wingo, D.: Fitting three parameter lognormal model by numerical global optimization-an improved algorithm, *Computational Statistics and Data Analysis*, No. 2 (1984), 13–25.

[62] Orsier, B. and Pellegrini, C.: Using global line searches for finding global minima of mlp error functions, In: *International Conference on Neural Networks and their Applications, Marseilles*, 1997, pp. 229–235.

[63] Groch, A., Vidigal, L. and Director, S.: A new global optimization method for electronic circuit design, *IEEE Trans. on Circuits and Systems* **32** (1985), 160–170.

[64] Perttunen, C., Jones, D. and Stuckman, B.: Lipshitzian optimization without the Lipshitz constant, *J. Optim. Theory Appl.* **79** (1993), 157–181.

Chapter 5

ANIMATED VISUAL ANALYSIS OF EXTREMAL PROBLEMS

Gintautas Dzemyda
Institute of Mathematics and Informatics
Akademijos St. 4, 2600 Vilnius, Lithuania
dzemyda@ktl.mii.lt

Abstract The results presented in this paper make up the basis for a new way of analyzing extremal problems. A new phenomenon that characterizes an extremal problem has been discovered. The paper tries to reveal fields of application of this phenomenon. The method of animated visual analysis, based on the knowledge discovery in the set of observations of the objective function of the problem interactively, has been developed. The aim of analysis is to find a direction in the definition domain such that maximizes the mean absolute difference between two values of the objective function calculated at randomly selected points in this direction, or (and) maximizes the mean absolute difference per distance unit of the objective function values calculated at two randomly selected points in this direction. The presented approach requires generating many data sets. Sometimes such a generation is very computation-expensive. Therefore, the ideas discussed in this paper may be applied in the case where the investigator wants not only to solve the extremal problem, but also to discover additional knowledge of it.

Keywords: Optimization, visual analysis, animation, data analysis, knowledge discovery

1. Introduction

Complex problems of computer-aided design and control arise with a rapid development of modern technologies. The search for optimal solutions acquires here an essential significance. Investigations in this area are pursued in two directions: development of new optimization methods as well as software that would embrace various realizations of the methods developed. The key aim is solution of any arising problem as soon as possible and with the least efforts. When selecting a proper strategy of optimization, the knowledge of the

G. Dzemyda et al. (eds.), Stochastic and Global Optimization, 65–91.
© 2002 *Kluwer Academic Publishers.*

extremal problem (the objective function, variables, and constraints) is of utmost importance. On the basis of the knowledge we choose a concrete method (or methods) for optimization. Most frequently, however, minimal knowledge of the problem to be optimized is used, i.e. as much as it is necessary for solving the problem (not necessarily in the most efficient way). With the view of improving the optimization efficiency, it is expedient to develop special methods and strategies for knowledge discovery about the optimization problem.

Knowledge discovery is the non-trivial extraction of implicit, previously unknown, and potentially useful information from data (see [1,2]). The term of knowledge discovery was introduced in [1] referring to the analysis of databases. The concept of knowledge discovery is extended to the analysis of information structures in [3].

By *information structure* we mean here a set of data (and knowledge, in certain cases) linked up in a certain way, the analysis of which can give us some new knowledge on an optimization problem and can help to solve the problem. The information structure may also cover other structures, simpler ones.

The main tasks in recent research are the search for information structures and their analysis strategies aimed at increasing the optimization efficiency and development of software that assists in the knowledge discovery.

Visualization is a powerful means to support optimization modeling, solution, and analysis. It uses interactive computer graphics to provide the insight of complicated extremal problems, models or systems. A broad survey on the use of visualization in optimization is given in [4]. Contrary to the static graphics, animation gives the illusion of motion. It can also help algorithm designers to understand the behavior of their algorithms, modelers their models, and decision makers their problems. The author in [4] touches the role of animation in the process of optimization, too. Two ways of application of the animation are reviewed in [4]: algorithm animation, i.e. demonstration of the behavior of execution of an algorithm (see for more details, e.g., [5] and [6]), and animated sensitivity analysis that provides the insight into the behavior of the solution produced by an algorithm (for more details, see, e.g., [7] and [8]). The review covers special algorithms and tools. However, other visualization approaches (e.g., projection approaches [9], worlds within worlds [10], parallel coordinates [11]) may also be adapted to animated usage seeking the goals of the abovementioned ways.

In this paper, we present a new way of visual and animated analysis of extremal problems. It allows us to discover new properties of the objective function and to search for a new coordinate system (i.e. for a linear transformation of the system of variables).

In fact, a new phenomenon that characterizes the extremal problem has been discovered. Its analysis, presented in the paper, raises more items for further discussions and research rather than gives answers. The paper tries to reveal fields of application of this phenomenon. However, a slight relationship with practical applications has been discovered.

2. The Aim of Analysis

Let us consider an extremal problem

$$\min_{X=(x_1,\ldots,x_n)\in\tilde{A}} f(X),$$

where $\tilde{A}$ is a bounded domain in an n-dimensional Euclidean space R^n, the objective function $f(X)$: $R^n \to R^1$ is continuous and multiextremal, in general. In particular, $\tilde{A}$ may be an n-dimensional rectangle $\tilde{A} = \{X : a_k \leqslant x_k \leqslant b_k, k = 1, \ldots, n\}$.

This paper is aimed at the analysis of the set of objective function values. Recent results of analysis of the set of objective function values are presented in [12–16]. In [13] and [14], a visual method of knowledge discovery in a set of objective function observations is proposed. The method makes it possible to find a direction, where the variation of the function is maximal. The results of the analysis are used in search of a new coordinate system of the extremal problem. The study in [13] and [14] seeks new features for the analysis of extremal problems. In the general case, the analysis allows us to base optimization on the information about the variation of function in various directions, relations among separate variables or their groups, and the structure of a calculation process of the function value. The proposed method is based on the visual analysis of graphically presented data: two-dimensional projections of n-dimensional data are analyzed.

In this paper, we extend the results of [13] and [14] by introducing the possibility of visual analysis of the animated data. We also review the main results of [14] and [15] that made a basis for the results of this paper and that are published in hardly accessible issues. This makes it clearer and easier readable. The main result of this paper is transfer of data that characterize the extremal problem to the level of animation.

The aim of the analysis is to find a direction in the definition domain $\tilde{A}$ such that

- maximizes the mean absolute difference between two values of the objective function calculated at randomly selected points in this direction (distribution of the points is uniform), or (and)
- maximizes the mean absolute difference per distance unit of the objective function values calculated at two randomly selected points in this direction (distribution of the points is uniform).

The quantities above are characteristics (not the only possible) of variation of the objective function. These characteristics are global for the definition domain $\tilde{A}$. Therefore, they may be useful in constructing the methods and strategies for global optimization and for the analysis of extremal problems. The main goal of the paper is to point out these characteristics and to draw preliminary conclusions on their usage. Further investigations should discover new advantages of these interesting characteristics.

In the general case, the directions optimizing both characteristics of variation are not identical, and the investigator should have a possibility to choose one of them or integrate both directions. Let us denote these directions by $\overline{Y_1}$ and $\overline{Y_2}$, respectively. The directions defined above may be useful in developing new optimization algorithms. If we start the minimization at a randomly selected point $X_{\mathrm{I}} \in \tilde{A}$, and execute a step of random length to the point $X_{\mathrm{II}} \in \tilde{A}$, then a mean absolute change in the objective function value will be maximal, in case both the points are in the first direction defined above. The second direction defined above (and the first direction, to some extent as well) is an extension of the gradient concept: the gradient $\nabla f(X)$ points to the direction of the steepest slope of the hypersurface of $f(\cdot)$ at the point X.

3. Information Structures

The initial information structure for the analysis is:

- points $X_i = (x_{i1}, \ldots, x_{in}) \in \tilde{A}$, $i = 1, \ldots, m$, $m \geqslant 2$, that compose a discrete set $D \subset \tilde{A}$,
- values (observations) of $f(X)$ at these points: $f(X_i)$, $i = 1, \ldots, m$.

The aim is to transform the initial information structure given above so that we could observe and analyze the variation of function in various directions. Two transformations η_1 and η_2 are proposed in [14]. These transformations are new (higher) information structures obtained on the basis of the initial information structure. The analysis of these new structures enables us to extract the knowledge on the extremal problem.

Seeking representative data sets for the analysis of the extremal problem, the points X_i, $i = 1, \ldots, m$, should cover the definition domain $\tilde{A}$ uniformly. These points should be selected from $\tilde{A}$ at random or in a certain deterministic manner.

Let any pair of points X_i, X_j, $i \neq j$, be taken from D with the same probability. In this case, we can observe the random n-dimensional quantities $\eta_1 = (\eta_{11}, \ldots, \eta_{1n})$ and $\eta_2 = (\eta_{21}, \ldots, \eta_{2n})$ whose values $\eta_s^{ij} = (\eta_{s1}^{ij}, \ldots, \eta_{sk}^{ij}, \ldots, \eta_{sn}^{ij})$, $s = 1, 2$, are uniquely related with a randomly selected pair X_i, X_j, $i \neq j$, as follows:

$$\eta_{1k}^{ij} = (|f^i - f^j|)^\tau \cdot \frac{x_{ik} - x_{jk}}{S_{ij}}, \qquad \eta_{2k}^{ij} = \left(\frac{|f^i - f^j|}{S_{ij}}\right)^\tau \cdot \frac{x_{ik} - x_{jk}}{S_{ij}},$$

where $f^i = f(X_i)$, $S_{ij} = \sqrt{\sum_{l=1}^{n} (x_{il} - x_{jl})^2}$, $\tau \in [-1, 1]$.

The transformations η_s, $s = 1, 2$, depend on τ. Therefore, in further formulae we sometimes use $\eta_s(0 < \tau < 1)$, $\eta_s(0.25)$, etc., where detailed values of τ are given in brackets.

Why did we introduce such transformations? Lengths of the vectors $\eta_s^{ij} = (\eta_{s1}^{ij}, \dots, \eta_{sn}^{ij})$, $s = 1, 2$, computed on the basis of a pair of randomly selected points X_i and X_j, are as follows:

$$\|\eta_1^{ij}\| = (|f^i - f^j|)^\tau, \qquad \|\eta_2^{ij}\| = \left(\frac{|f^i - f^j|}{S_{ij}}\right)^\tau.$$

Therefore, if τ is positive, then a longer distance of point η_s^{ij} from the center $(0, 0, \dots, 0)$ corresponds to a greater 'variation of function' in the direction defined by the pair of points X_i and X_j. The longer distance corresponds to a smaller 'variation of function' in the case of negative τ.

We have an n-dimensional argument $X = (x_1, \dots, x_k, \dots, x_n)$ of the objective function $f(X)$: $R^n \to R^1$ and two n-dimensional quantities $\eta_s = (\eta_{s1}, \dots, \eta_{sk}, \dots, \eta_{sn})$, $s = 1, 2$. In fact, the kth component η_{sk} of both quantities η_1 and η_2 is related with the kth variable x_k of the extremal problem. So, a pair of components (say, η_{sk} and η_{sl}) of the n-dimensional quantities η_1 or η_2 will describe a relationship between two corresponding variables x_k and x_l.

Examples of distributions of the values of η_1 and η_2 are presented graphically in Figures 1–4 for four functions dependent on two variables ($m = 30$, $n = 2$, $\eta_1 = (\eta_{11}, \eta_{12})$, $\eta_2 = (\eta_{21}, \eta_{22})$):

1. Linear function: $f_1 = x_1 + 3x_2$, $-1 \leqslant x_1, x_2 \leqslant 1$.
2. Piecewise linear function: $f_2 = |x_1 - 3x_2| + x_1 + 3x_2$, $-1 \leqslant x_1, x_2 \leqslant 1$.
3. Quadratic function: $f_3 = (x_1 - x_2)^2 + [(x_1 + x_2)/2]^2$, $-1 \leqslant x_1, x_2 \leqslant 1$.
4. Multiextremal Branin's function (three local minima) [17]:
   $f_4 = (x_2 - 0.1292x_1^2 + 1.59155x_1 - 6)^2 + 9.60211\cos(x_1) + 10$,
   $-5 \leqslant x_1 \leqslant 10$, $0 \leqslant x_2 \leqslant 15$.

The data presentation in Figures 1–4 needs additional comment.

$m = 30$ points $X_i = (x_{i1}, x_{i2})$, $i = 1, \dots, m$, were generated randomly in the definition domain $\tilde{A}$ and the values $f(X_i)$, $i = 1, \dots, m$, of $f(X)$ were computed at these points. The values $\eta_s^{ij} = (\eta_{s1}^{ij}, \eta_{s2}^{ij})$, $i, j = 1, \dots, m$, $i \neq j$, of $\eta_s = (\eta_{s1}, \eta_{s2})$, $s = 1, 2$, were computed on the basis of all possible pairs of $X_i = (x_{i1}, x_{i2})$, $i = 1, \dots, m$.

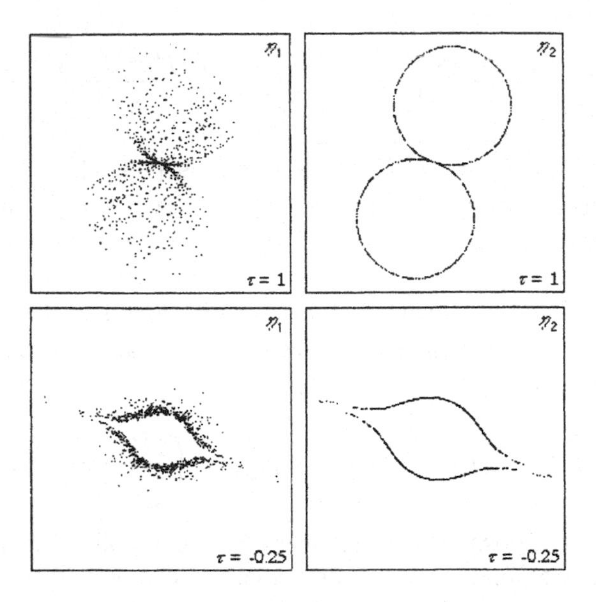

Figure 1. Distributions for f_1.

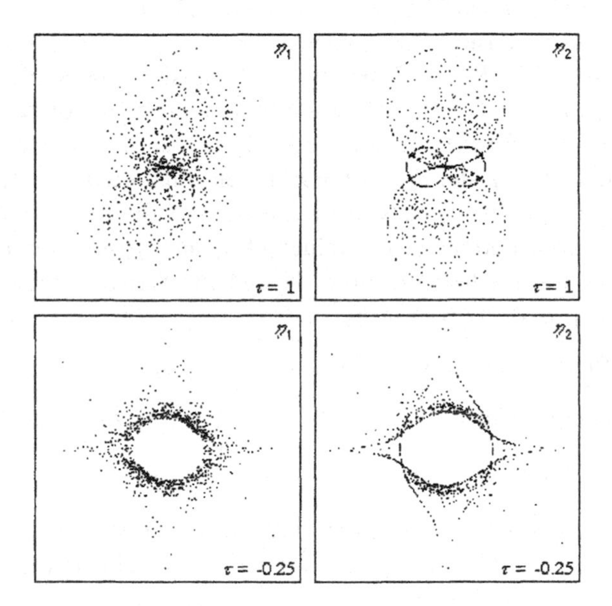

Figure 2. Distributions for f_2.

Each figure consists of four pictures. Each picture represents the distribution of the values of $\eta_1 = (\eta_{11}, \eta_{12})$ or $\eta_2 = (\eta_{21}, \eta_{22})$, i.e. cases $s = 1$ and $s = 2$, for two different values of $\tau (\tau = 1$ or $\tau = -0.25)$. The abscissa is destined for the values of η_{s1}, and the ordinate is destined for the values of η_{s2}.

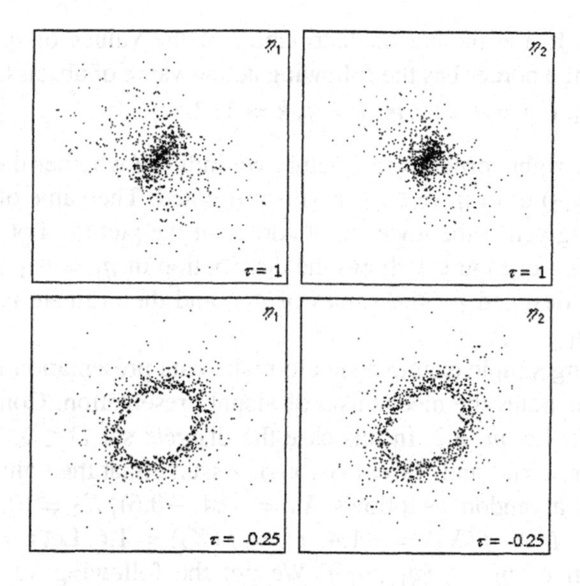

Figure 3. Distributions for f_3.

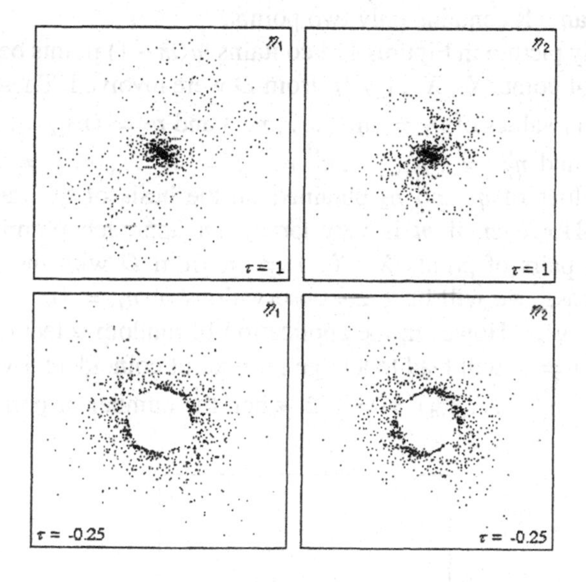

Figure 4. Distributions for f_4.

The data are scaled in the pictures:

- The center of the picture corresponds to $\eta_s = (0, 0)$.
- The shortest distance from a border of any picture to the nearest point of the distribution, presented in this picture, is the same through all the

pictures. In the picture of distribution of the values of η_s, the nearest point to the border has the following actual value of abscissa or ordinate: $\max|\eta_{sk}^{ij}|$, $i, j = 1, \ldots, m, i \neq j, k = 1, 2$.

At the upper right corner of the picture we indicate whether the distribution of $\eta_1 = (\eta_{11}, \eta_{12})$ or of $\eta_2 = (\eta_{21}, \eta_{22})$ is presented. The value of τ ($\tau = 1$ or $\tau = -0.25$) is given at the lower right corner of the picture. For example, the upper left picture of Figure 1 shows the distribution of $\eta_1 = (\eta_{11}, \eta_{12})$; $\tau = 1$; the abscissa is destined for the values of η_{11}, and the ordinate is destined for the values of η_{12}.

The following simple example should make data presentation in Figures 1–4 clearer. It illustrates the mechanism of visual presentation. Consider the linear function f_1. Let $m = 2$. In this case the discrete set $D \subset \tilde{A}$ contains two points $X_1 = (x_{11}, x_{12})$ and $X_2 = (x_{21}, x_{22})$. Assume that the points X_1 and X_2 were generated at random as follows: $X_1 = (0.4, -0.6)$, $X_2 = (0.4, 0.4)$. Then $S_{12} = S_{21} = 1$, $f^1 = f(X_1) = -1.4$, $f^2 = f(X_2) = 1.6$. Let $\tau = 1$. Consider the distribution of $\eta_1 = (\eta_{11}, \eta_{12})$. We get the following values of $\eta_1 = (\eta_{11}, \eta_{12})$: $\eta_1^{12} = (0, -1)$ and $\eta_1^{21} = (0, 1)$. Figure 5 shows the distribution of η_1 on the plane. It contains only two points.

Note that any picture in Figures 1–4 contains $m(m-1)$ points because all the possible pairs of points $X_i, X_j, i \neq j$, from D were involved. These points produced $m(m-1)$ values of $\eta_1 = (\eta_{11}, \ldots, \eta_{1n})$ and $\eta_2 = (\eta_{21}, \ldots, \eta_{2n})$: $\eta_1^{ij} = (\eta_{11}^{ij}, \ldots, \eta_{1n}^{ij})$ and $\eta_2^{ij} = (\eta_{21}^{ij}, \ldots, \eta_{2n}^{ij})$, $i, j = 1, \ldots, m, i \neq j$. These are all possible values of η_1 and η_2 obtained on the basis of the randomly generated set D. However, if m is very large, our approach permits a random selection of ω pairs of points $X_i, X_j, i \neq j$, from D with the same probability. In this case, we will have 2ω observations of $\eta_1 = (\eta_{11}, \ldots, \eta_{1n})$ and $\eta_2 = (\eta_{21}, \ldots, \eta_{2n})$. However, the application of random selection of pairs of points $X_i, X_j, i \neq j$, will lead to a large number of coincidental values among $\eta_s^{ij} = (\eta_{s1}^{ij}, \ldots, \eta_{sk}^{ij}, \ldots, \eta_{sn}^{ij})$, $s = 1, 2$, when the number of points m in D is

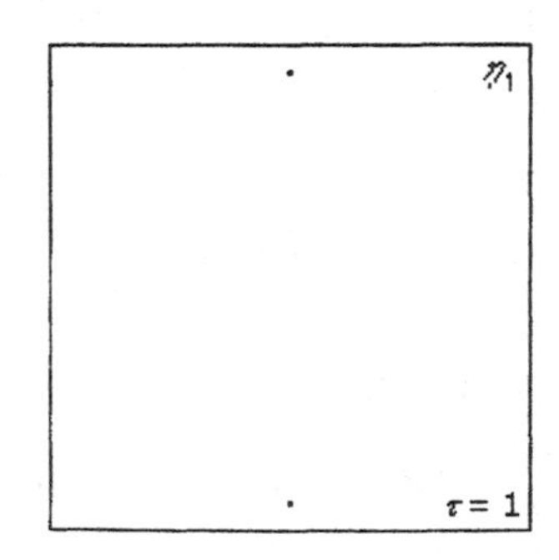

Figure 5. Illustration of the mechanism of visual presentation of distributions.

small. In all the further experiments, we compute the distributions of η_1 and η_2 for all possible pairs of points $X_i, X_j, i \neq j$, from D.

The regular forms of distributions given in Figures 1–4 show that their analysis should lead to the detection of new peculiarities of the objective function. From the figures we can derive three criteria for their visual analysis:

C1: direction, in which the density of points is maximal.

C2: global orientation of distribution on the plane.

C3: special regularities.

Let us comment these criteria.

C1. Search for direction, in which the density of points is maximal, allows us to estimate $\overline{Y_1}$ and $\overline{Y_2}$: the analysis of distributions of the values of η_1 in the case of positive τ allows us to search for a direction $\overline{Y_1}$ in the definition domain $\tilde{A}$ that maximizes the mean absolute difference between two values of the objective function calculated at randomly selected points in this direction, while the analysis of distributions of the values of η_2 in the case of positive τ allows us to search for a direction $\overline{Y_2}$ that maximizes the mean absolute difference per distance unit of the objective function values calculated at two randomly selected points in this direction.

C2. We observe in Figures 1, 3, and 4 the exactly expressed global orientation of distributions on the plane. By the orientation of distribution we call here the degree of rotation of a totality of points in the picture. A unit vector starting from the center of the picture may describe the orientation. For example, $f_1(X) = x_1 + 3x_2$ is a linear function, and from Figure 1 we can intuitively observe that the orientation of distributions both of $\eta_1 = (\eta_{11}, \eta_{12})$ and $\eta_2 = (\eta_{21}, \eta_{22})$ for the case $\tau = 1$ is defined by a unit vector $(1/\sqrt{3}, 3/\sqrt{3})$. For $\tau = -1$ the orientation is perpendicular to the direction above.

C3. Special regularities can be observed in Figure 2 – see distributions of $\eta_2 = (\eta_{21}, \eta_{22})$ for the cases $\tau = 1$ and $\tau = -1$. Both the pictures (distributions of $\eta_2(1)$ and $\eta_2(-1)$) look like consisting of two separate distributions that are put on the same plane. The reason for such strange and visually interesting distributions is in the structure of the piecewise linear function f_2. The regularities can be observed in Figure 1 for distributions of $\eta_2(1)$ and $\eta_2(-1)$, too. From Figures 1 and 2 we observe a tendency that linearity in the objective function $f(X)$ produces the appearance of figure eight in the picture of distribution of $\eta_2(1)$.

In this paper, we put a stress on search for advantages of the first criterion C1 – i.e. of the directions $\overline{Y_1}$ and $\overline{Y_2}$. As stated in the section above, we can draw some parallels between the directions found in $\tilde{A}$ and the gradient. However, the gradient depends on a point in $\tilde{A}$, and the new directions are global for the whole definition domain $\tilde{A}$: they represent a general variation of hypersurface of $f(X)$ in $\tilde{A}$. Decisions referring to criterion C2 can be made analyzing covariance matrices of $\eta_1 = (\eta_{11}, \ldots, \eta_{1n})$ and $\eta_2 = (\eta_{21}, \ldots, \eta_{2n})$ – see Sections 4.1 and 5. Special regularities (criterion C3) are discussed in Section 8, too.

It follows from Figures 1–4 that a lot of points concentrate in the center of pictures in case $|\tau| = 1$. It means that, in case $|\tau| = 1$, a visual decision may often be influenced by a significantly smaller number of points located near the border of pictures. A natural problem arises: how to present the points that are located in the center of the picture to the investigator? This may be done by varying the value of τ. The dependence of distributions of the values of η_1 and η_2 on τ is illustrated in Figure 6 for the linear function f_1 and the quadratic function f_3, $\tau = 1, 0.25, 0, -0.25, -1$. Let us compare three pairs of distributions: $\eta_1(1)$ and $\eta_1(0.25)$, $\eta_1(-1)$ and $\eta_1(-0.25)$, $\eta_2(-1)$ and $\eta_2(-0.25)$. In all the cases, if $|\tau| = 1$ we observe some concentration of points in the centers of pictures. When $|\tau| < 1$, the centers of pictures become empty and we can observe some empty regular forms in the centers of pictures. These regularities may give an additional information for human decision. Dependencies in Figure 6 allow us to conclude that varying τ one can analyze functions better, i.e. often it is better to use $0 < \tau < 1$ and $-1 < \tau < 0$ instead of $\tau = 1$ and $\tau = -1$. For the advantages of usage of different values of τ see also Section 8.

4. Analysis

Three possibilities of directional analysis are discussed below:

1. The factor analysis of covariance matrices of $\eta_s = (\eta_{s1}, \ldots, \eta_{sn})$, $s = 1, 2$.
2. The interactive visual analysis of observations of η_s, $s = 1, 2$.
3. Combination of the visual and factor analysis.

4.1. Application of the Factor Analysis

The factor analysis [20] is a classical method for analysis of correlation and covariance matrices. It enables us to find a linear combination $\vartheta_1^s = a_{11}^s\eta_{s1} + a_{12}^s\eta_{s2} + \cdots + a_{1n}^s\eta_{sn}$ of $\eta_{s1}, \ldots, \eta_{sn}$, $s = 1, 2$, having the maximal variance. The solution $a_1^s = (a_{11}^s, \ldots, a_{1n}^s)$ is a normalized eigen-vector (i.e. a vector of unit length) corresponding to the greatest eigen-value of the covariance matrix $K_s = \{K_{\eta_{sk}\eta_{sl}}, \ k, l = 1, \ldots, n\}$ of η_s. Therefore, applying the factor analysis in the case $\tau > 0$, the criterion of 'variation of function' may be the greatest eigen-value of the covariance matrix. The eigen-vector corresponding to the eigen-value represents the desired direction. The factor analysis may also be used in the case $\tau < 0$, but the first found direction $a_1^s = (a_{11}^s, \ldots, a_{1n}^s)$ will show where the 'variation of function' is the least one.

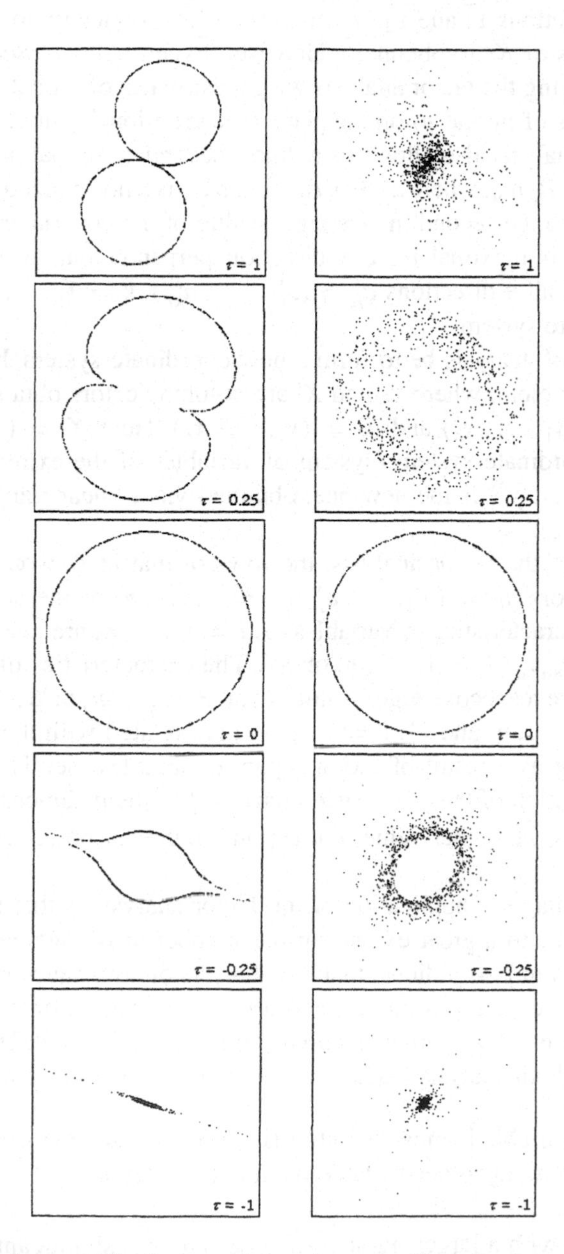

Figure 6. Dependences of distributions of η_1 and η_2 on τ. The left column – the distribution of η_2 for the linear function f_1; the right column – the distribution of η_1 for the quadratic function f_3.

At this point we would be able to conclude that the factor analysis allows us to estimate directions $\overline{Y_1}$ and $\overline{Y_2}$: the analysis of covariance matrix K_1 to obtain $\overline{Y_1}$, and analysis of K_2 to obtain $\overline{Y_2}$. However, experiments in Section 5 refute this wish: applying the factor analysis we get estimates of global orientation of the distributions of η_1 and η_2 on the plane (see criterion C2 in Section 3).

The factor analysis also allows us to find a second combination $\vartheta_2^s = a_{21}^s \eta_{s1} + a_{22}^s \eta_{s2} + \cdots + a_{2n}^s \eta_{sn}$, where $a_2^s = (a_{21}^s, \ldots, a_{2n}^s)$ is a normalized eigen-vector, corresponding to the second in size eigen-value of the covariance matrix. The variance of ϑ_2^s is maximal for any direction perpendicular to a_1^s. Thus, it is possible to find all n directions $a_k^s = (a_{k1}^s, \ldots, a_{kn}^s)$, $k = 1, \ldots, n$, that create a new coordinate system.

Let the aim of analysis be to form a new coordinate system $Y' = UX'$ for the extremal problem, where Y' and X' are column vectors obtained from row vectors $Y = (y_1, \ldots, y_n)$ and $X = (x_1, \ldots, x_n)$. Here $X = (x_1, \ldots, x_n)$ is the present coordinate system (system of variables of the extremal problem) and $Y = (y_1, \ldots, y_n)$ is the new one, obtained via a linear transformation of $X = (x_1, \ldots, x_n)$.

As a result of the factor analysis, the rows of matrix U would be normalized eigen-vectors $a_i^s = (a_{i1}^s, \ldots, a_{in}^s)$, $i = 1, \ldots, n$, of the covariance matrix K_s. The characteristics of variables x_i, $i = 1, \ldots, n$, are respective diagonal elements $K_{\eta_{sk}\eta_{sk}}$, $k = 1, \ldots, n$, of K_s. The characteristics of variables y_i, $i = 1, \ldots, n$, are respective eigen-values λ_i^s, $i = 1, \ldots, n$, of K_s. The values of $K_{\eta_{sk}\eta_{sk}}$, $k = 1, \ldots, n$, and λ_i^s, $i = 1, \ldots, n$, are related with the optimization errors occurring as a result of fixing separate variables (see [15] for experimental investigation of the covariance matrix K_1^* of the n-dimensional random quantity $\eta_1(1)/\sqrt{2}$): higher values correspond to greater possible optimization errors.

The disadvantage of application of the factor analysis is that the quality of analysis depends, to a great extent, on the number m of calculated values of $f(\cdot)$. The results in [14] indicate that the analysis of covariance matrices gives poor results in the case of small m, and a new observation of $f(\cdot)$ or another set of observations of $f(\cdot)$ may essentially influence the result. However, even for small m such an analysis leads to a new coordinate system, where

- the first variable is most essential (its fixing causes the greatest error as compared to any other variable from both systems),
- variables with a larger number of order are less significant (their fixing causes smaller errors), and the variables with the largest numbers of order are often less significant than any variable from the old system,
- coordinate descent gives better results.

4.2. Interactive Visual Analysis

The main idea of such an analysis is to present the sets of observations of $\eta_s = (\eta_{s1}, \ldots, \eta_{sn})$, $s = 1, 2$, to the investigator graphically. The investigator makes a decision on the best direction. Therefore, the main direction (not the system of perpendicular directions) may be found in this manner only.

The algorithm of visual analysis:

1. The pictures of distributions of (η_{sk}, η_{sl}) are presented graphically to the investigator for any pair (k, l) $k, l = 1, \ldots, n$, $k < l$, of variables in consecutive order (the values of $(\eta_{sk}^{ij}, \eta_{sl}^{ij})$ are presented in these pictures – see Section 3).

2. The investigator analyses the distributions visually and shows the best, to his mind, direction $\alpha_{kl}x_k - \alpha_{lk}x_l = 0$ for any pair (k, l), $k, l = 1, \ldots, n$, $k < l$, of variables. Here a_{kl} is the coefficient at x_k when x_k is in the same equation as x_l. The decision depends on the goal of analysis and may be made on the basis of distributions of separate random quantities or groups of distributions (e.g., quantities η_s, $s = \overline{1, 2}$, for various values of τ).

3. The integral direction is determined by $n(n-1)/2$ subdirections

$$\alpha_{kl}x_k - \alpha_{lk}x_l = 0, \quad k, l = 1, \ldots, n, \; k < l. \tag{1}$$

Such an interactive visual analysis may be used to determine directions described by criterion C1 (directions $\overline{Y_1}$ and $\overline{Y_2}$) or by criterion C2 (see Section 3).

Sometimes it may be useful to analyze the distributions of η_1 and η_2 at the same time because the directions $\overline{Y_1}$ and $\overline{Y_2}$ are frequently similar (see Figures 7 and 8). Our experience shows that it would be better to present to the investigator the distributions of $\eta_1(\tau < 0)$ and $\eta_2(\tau < 0)$ rotated by the 90° angle, if he prefers to analyze the distributions of η_1 and η_2 for both positive and negative values of τ at the same time.

The system of equations (1) consists of $n(n-1)/2$ equations. Each equation means a subdirection in R^n. We present below the approach for finding the integral direction on a basis of the system of subdirections (1).

Problems of Finding the Integral Direction. The direction in the n-dimensional Euclidean space R^n is entirely defined if the coordinates of start and end points of the direction vector are known. The start point in our case is $X^0 = (0, \ldots, 0)$. The end point $X^* = (x_1^*, \ldots, x_n^*)$ must be determined from the system of $n(n-1)/2$ equations (1). Let the distance between X^0 and X^* be equal to 1. Then the problem is as follows:

$$a_{kl}x_k - a_{lk}x_l = 0, \quad k, l = 1, \ldots, n, \; k < l, \tag{2}$$

$$\sum_{l=1}^{n} x_l^2 = 1. \tag{3}$$

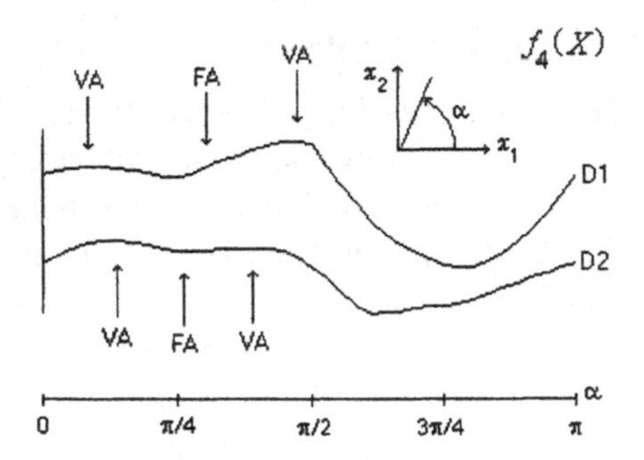

Figure 7. Dependence of D1 and D2 on the angle α.

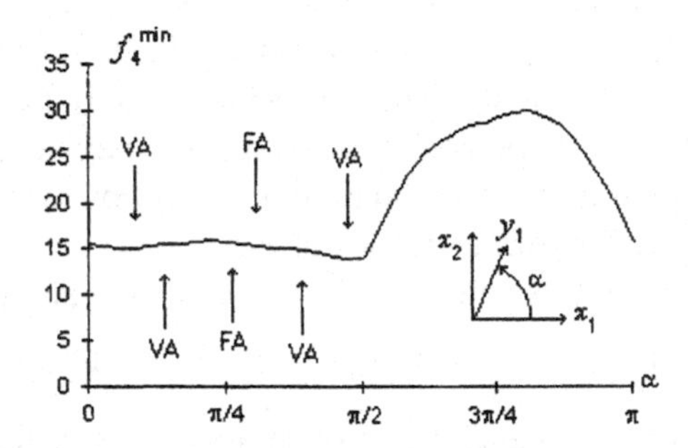

Figure 8. Averaged results of minimization.

The number of equations in the system of linear equations (2) starting from $n > 2$ is larger than the number of variables. However, it is necessary to find a solution. In this case, problem (2)–(3) may be formulated and solved as an optimization one using the least squares approach. After some transformations, problem (2)–(3) looks like this (see proof in [14]):

$$\min_{\substack{x_i \in [-1,1], i=1,\dots,n \\ X \neq (0,0,\dots,0)}} \gamma^*(x_1, \dots, x_n)$$

$$= \sum_{k=1}^{n} \left(\sum_{\substack{l=1 \\ l \neq k}}^{n} (a_{kl}x_k - a_{lk}x_l)a_{kl} - \varsigma x_k \right)^2 + \left(\sum_{l=1}^{n} (x_l)^2 - 1 \right)^2 + \beta(\lambda_{\min} - \varsigma)^2,$$

where $\lambda_{\min}$ is the least eigen-value of the matrix

$$B = \left\{ b_{kk} = \sum_{i=1, i \neq k}^{n} a_{ki}^2,\ b_{kl} = -a_{lk}a_{kl},\ k, l = 1, \dots, n, l \neq k \right\},$$

$$\varsigma = \frac{\sum_{k=l}^{n} x_k \sum_{l=1 l \neq k}^{n} (a_{kl}x_k - a_{lk}x_l)a_{kl}}{\sum_{k=1}^{n} x_k^2}, \qquad \beta = \begin{cases} 0 & \text{if } \varsigma < \lambda_{\min}, \\ 1 & \text{if } \varsigma \geqslant \lambda_{\min}. \end{cases}$$

The problem above may be solved using an algorithm for the constrained optimization. The starting point for local search may be $x_i = 1/\sqrt{n}, i = 1, \dots, n$. The variable metric algorithm from MINIMUM [18] gave good results in minimizing the function $\gamma^*(x_1, x_2, \dots, x_n)$.

4.3. Combination of the Visual and Factor Analysis

Various combinations of the visual and factor analysis are possible. For example:

1. Factor analysis precedes the visual analysis: determination of direction using the factor analysis; an interactive visual analysis taking into account the results of the factor analysis.

2. Visual analysis precedes the factor analysis: an interactive visual analysis taking into account the results of the factor analysis; creation of a new coordinate system by finding other $n - 1$ directions using the factor analysis.

3. Combination of the previous two strategies: determination of direction using the factor analysis; an interactive visual analysis taking into account the results of the factor analysis; creation of a new coordinate system by finding other $n - 1$ directions using the factor analysis.

In the first case, the results of the factor analysis serve as initial data for the visual analysis. The aim of analysis in the second case is to make up a new coordinate system $Y' = UX'$ taking into account the results of the visual analysis. As a result of the visual analysis, let the direction $a = (\bar{a}_1, \dots, \bar{a}_n)$ be found. $\|a\| = 1$. It is proposed in [14] to modify the covariance matrix K_s by adding a new matrix $K_a = \{\lambda \bar{a}_i \bar{a}_j, i, j = 1, \dots, n\}$, where $\lambda > 0$. A further examination of the matrix $K_a + K_s$ is performed using the factor analysis.

5. Examples of Application of the Visual Analysis and Its Relationship with the Results of Optimization

Experiments were carried out on various test functions. We analyze here the results regarding f_4. The mean absolute difference (D1) and the mean absolute difference per distance unit (D2) for this function calculated at randomly selected pairs of points in the direction whose orientation defines the angle α are presented in Figure 7. Distributions of the values of $\eta_1(1)$ and $\eta_2(1)$ were investigated by the factor analysis (FA) and the visual analysis (VA).

It follows from Figure 7 that both D1 and D2 have two maxima for f_4. The visual analysis, using different data sets that contain 30 randomly selected points in $\tilde{A}$ ($m = 30$), indicated both maxima by different respondents: decisions made by the respondents are located around the pointers denoted as VA. The factor analysis yielded a direction that is an attempt to integrate both maxima of D1 and D2. In most cases of the analysis of distributions of the values of $\eta_1(1)$ and $\eta_2(1)$, the factor analysis indicated directions other than those corresponding to the maxima of D1 and D2. Indeed, the factor analysis showed the global orientation of distributions of the values of $\eta_s(1)$ on the plane (see criterion C2 in Section 3). The visual analysis of distributions of the values of $\eta_1(-0.25)$ and $\eta_2(-0.25)$ gave similar results to that of the factor analysis. It means that the visual analysis of distributions of the values of $\eta_1(\tau < 0)$ and $\eta_2(\tau < 0)$ also makes it possible to estimate the global orientation of distributions of the values of $\eta_1(\tau > 0)$ and $\eta_2(\tau > 0)$, respectively.

The possibility and efficiency of applying the proposed visual analysis in the optimization is illustrated in Figure 8. The extremal problem, containing Branin's function $f_4(x_1, x_2)$, was analyzed. The problem was simplified by introducing a new coordinate system $Y = (y_1, y_2)$ and by fixing the variable y_2. Five steps of random search for the minimum of the function, dependent on a single variable, have been made. The results were averaged by 100000 searches. The averaged minimal value $f_4^{\min}$ is presented in Figure 8 dependent on α. The pointers repeat the results of the visual and factor analysis from Figure 7.

The efficiency of visual analysis depends on the software abilities. The software with a 'mouse'-managed choice of the best direction has been developed. It simplifies the visual analysis (see Figure 9).

6. Background for the Animated Visual Analysis

The dependence of distribution of the values of η_1 and η_2 on the set D of the points $X_i = (x_{i1}, \ldots, x_{in})$, $i = 1, \ldots, m$, was discovered in [14]. The experiments were carried out on the problem of computer-aided synthesis of the external circuit of a tunable subnanosecond pulse TRAPATT-generator [19]. The sets of values of η_1 were investigated. The number m of points $X_i = (x_{i1}, \ldots, x_{in})$, $i = 1, \ldots, m$, was selected equal to 70. The values of $x_3, \ldots, x_7$ were fixed and equal to 0, and the values of x_1 and x_2 were varied in [0,1]. In Figure 10, the distributions of $\eta_1(1)$ for eight different sets of points X_i,

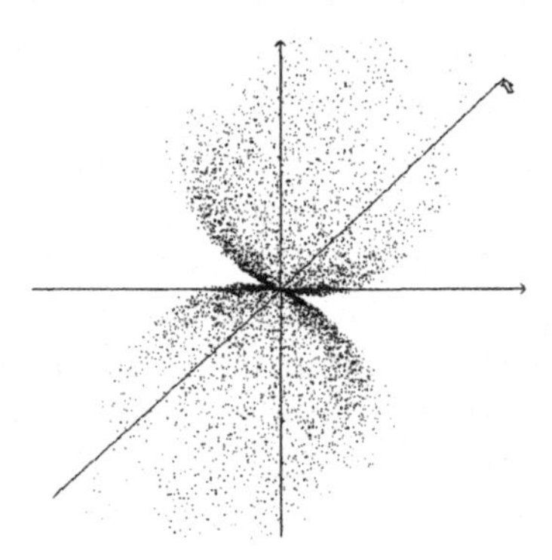

Figure 9. Example of the process of visual analysis.

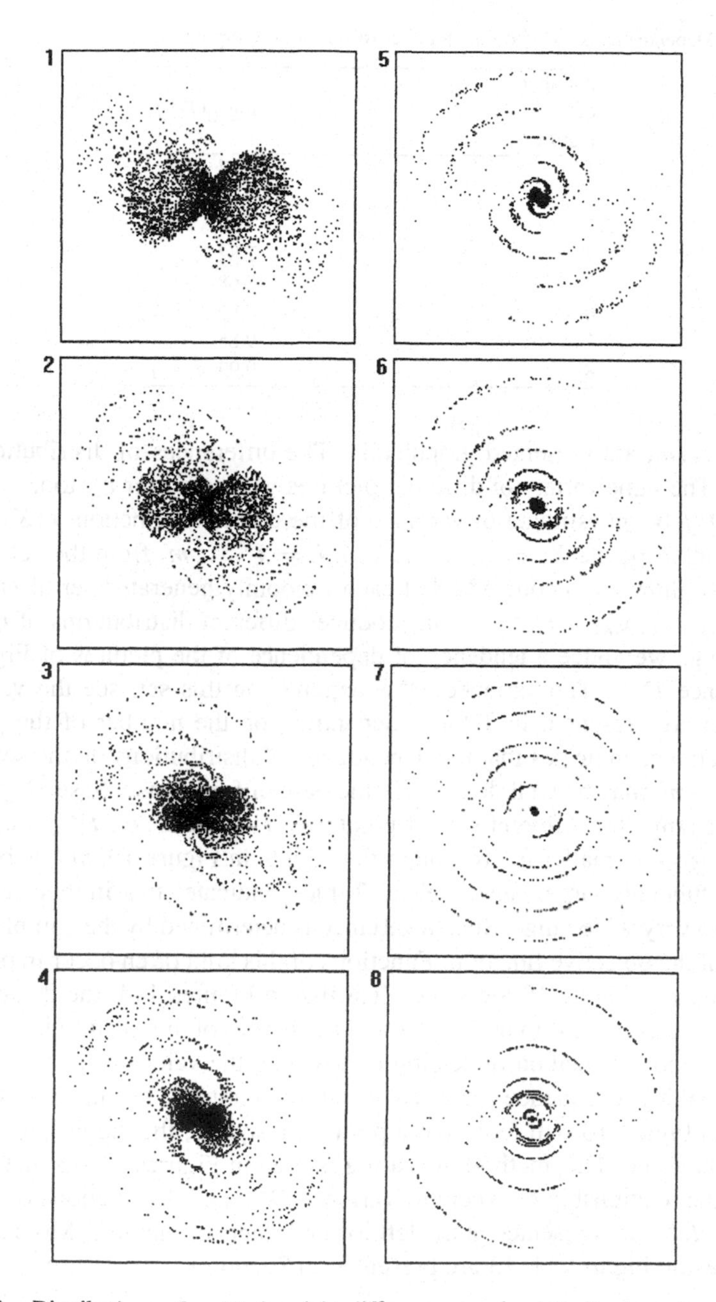

Figure 10. Distributions of $\eta_1(1)$ for eight different sets of points X_i, $i = 1, \ldots, m$; the objective function is computer-aided synthesis of the external circuit of a tunable subnanosecond pulse TRAPATT-generator.

Table 1. Dependence of D^*f on the number of pictures in Figure 10

Number of picture	$\log_{10}(D^*f)$
1	7.07
2	7.25
3	7.37
4	7.82
5	8.68
6	8.82
7	9.85
8	9.92

$i = 1, \dots, m$, are presented graphically. The differences of distributions are evident. The same variety of different pictures was noted for η_2, too.

Let D^*f be an estimate of variance of the objective function $f(X)$ on the basis of all m points $X_i = (x_{i1}, \dots, x_{in})$, $i = 1, \dots, m$, from the set D. Experiments allow us to conclude that each randomly generated set of m points $X_i = (x_{i1}, \dots, x_{in})$, $i = 1, \dots, m$, produces different distributions of η_1 or η_2 values, e.g., we notice a tendency of dependence of the pictures of Figure 10 on variance D^*f. The variance D^*f depends on this set: see the values of D^*f that are presented in Table 1 depending on the number of the picture. Our experience indicates that the dependence of distributions on the set of X_i, $i = 1, \dots, m$, may be well detected in the case of functions whose D^*f varies in a wide range for different sets of points X_i, $i = 1, \dots, m$. D^*f varies in a wider range for smaller m. To obtain the results in Figure 10, m has been selected sufficiently large. Nevertheless, Table 1 indicates that in this case D^*f varies in a very wide range. Such a variance is determined by the complexity of surface of the objective function. Functions of this kind often occur in practice.

The data in Figure 10 are scaled just like in Figures 1–4: the distance between the nearest point to the border and the border of any picture is the same. However, another method of scaling is possible: the values of borders of any picture making a sequence are fixed for all the pictures. In this case, the first picture in Figure 10 will look like a point in a square; the eighth picture will remain the same. This method of scaling is used in Figures 11–13 that also illustrate the relationship between the variance D^*f and distributions of η_1. The values of D^*f are presented at the left lower corners of pictures. More detailed comments on Figures 11–13 are presented in Section 8.

7. Animation Algorithm

Different distributions of η_1 or η_2 values may influence the decision of an investigator in a contradictory way. It means that in most cases a single dis-

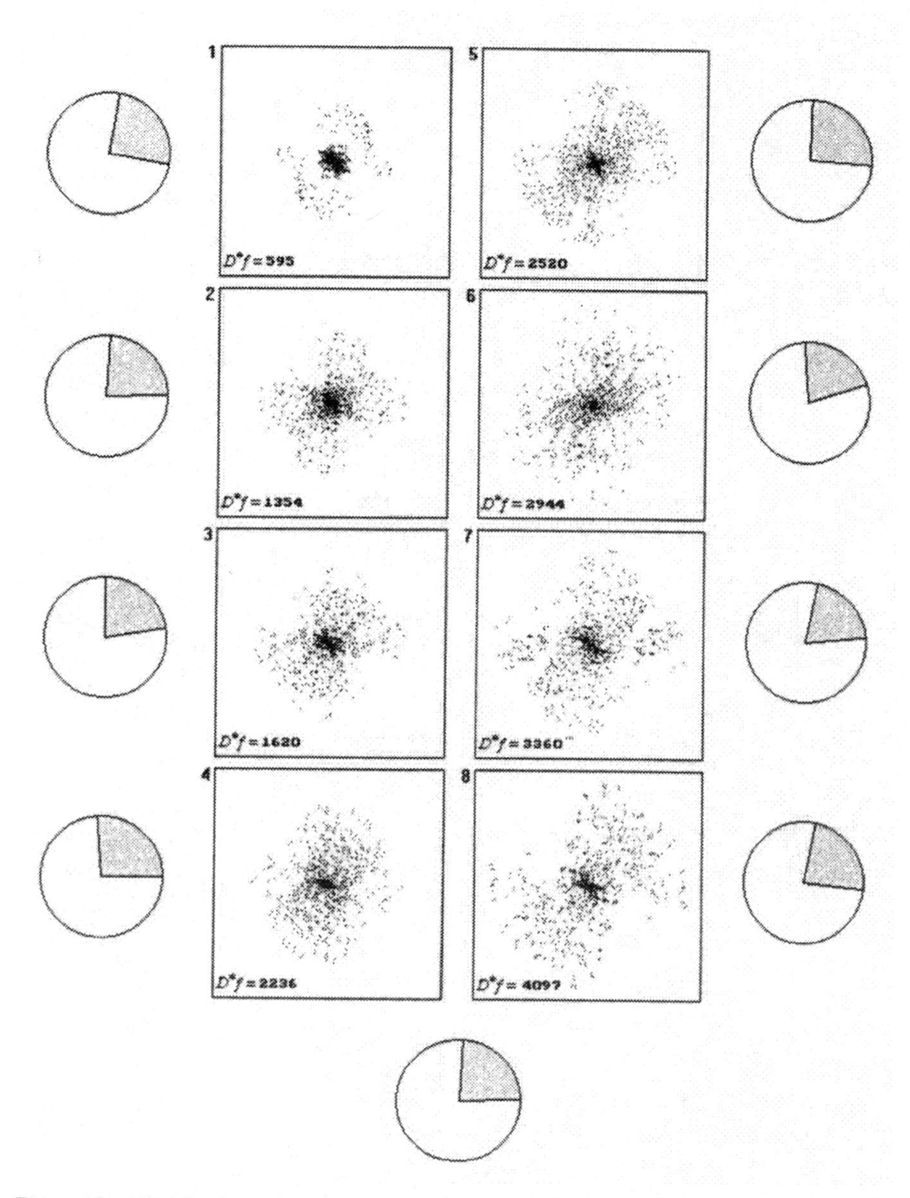

Figure 11. Distributions of $\eta_1(1)$ for eight different sets of points X_i, $i = 1, \ldots, m$; $m = 30$, $v = 50$, $p = 8$, Branin's function f_4.

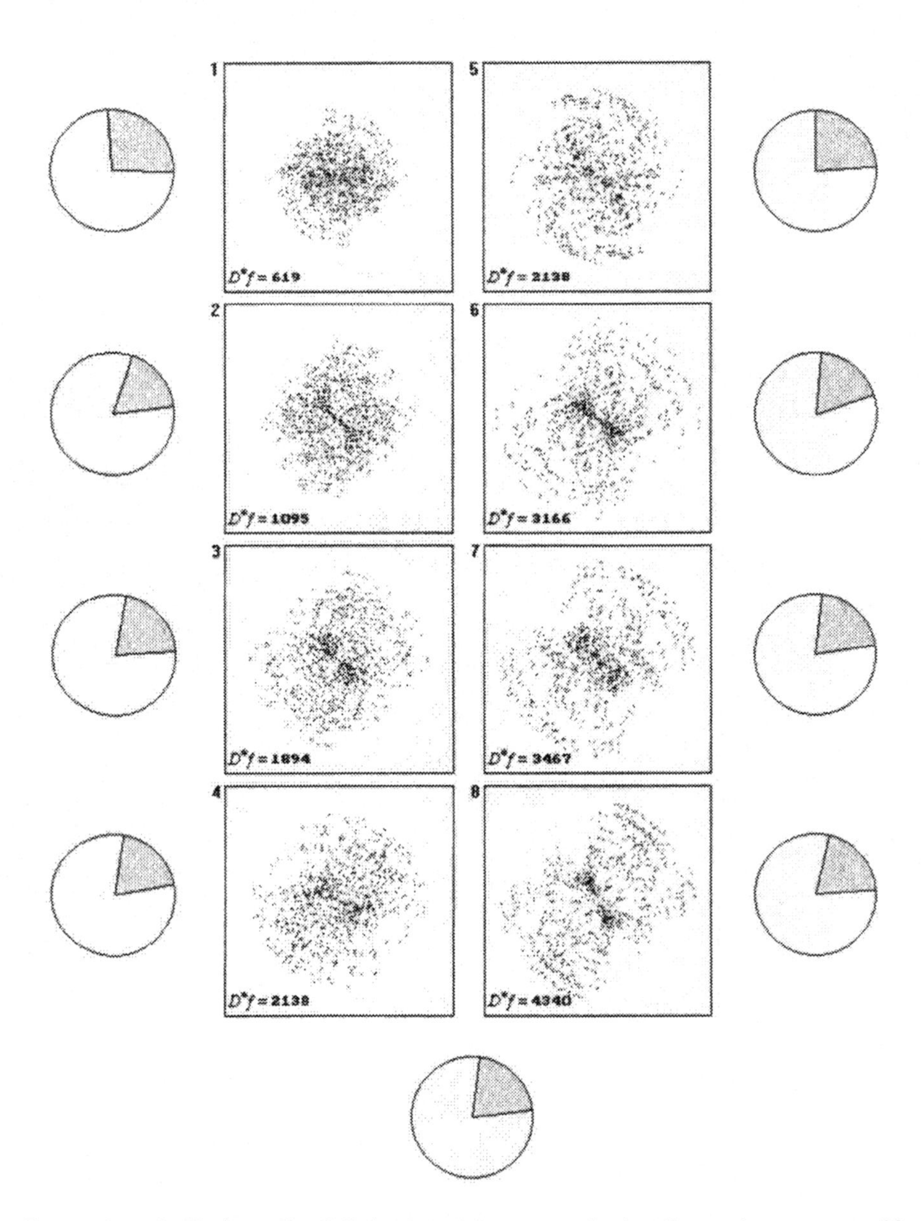

Figure 12. Distributions of $\eta_1(0.5)$ for eight different sets of points X_i, $i = 1, \ldots, m$; $m = 30$, $v = 50$, $p = 8$, Branin's function f_4.

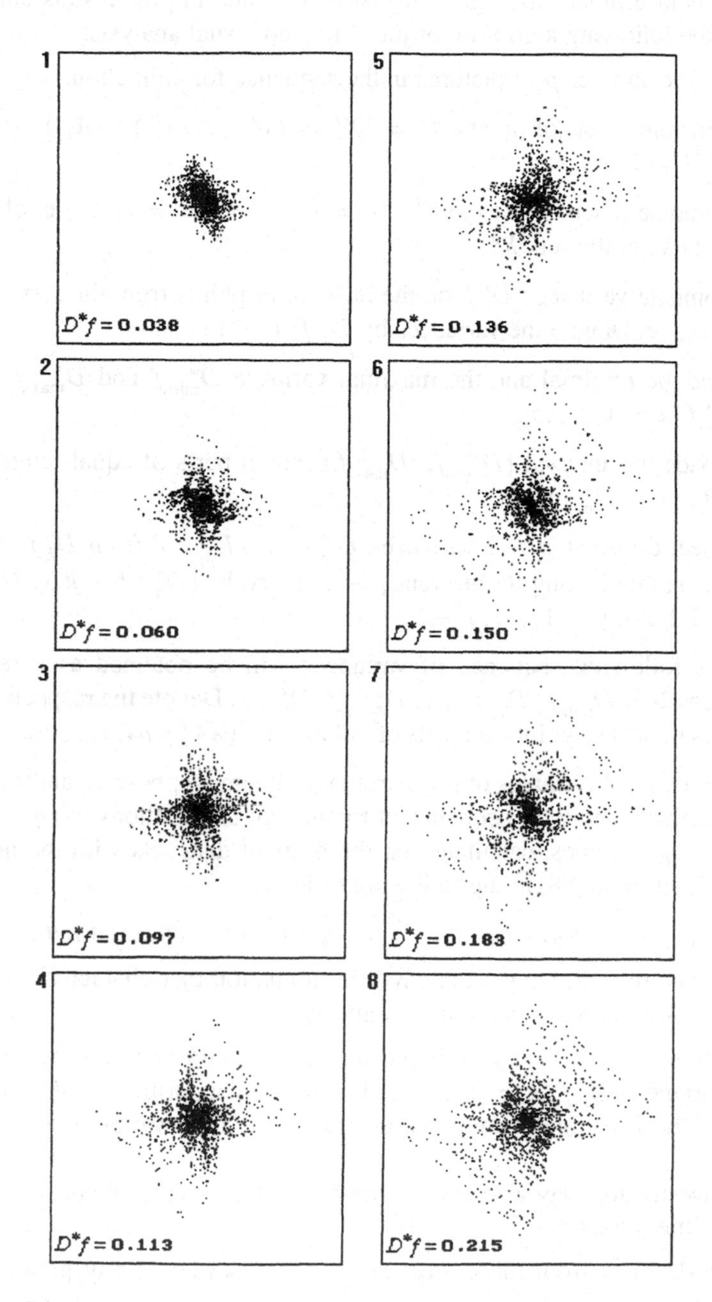

Figure 13. Distributions of $\eta_1(1)$ for eight different sets of points X_i, $i = 1, \ldots, m$; $m = 40$, $v = 40$, $p = 8$, the objective function $f_5 = \exp(x_1 x_2)$.

tribution is insufficient for a good decision. Considering the results above we propose the following algorithm of the animated visual analysis:

1. Fix the number p of pictures in the sequence for animation.
2. Generate v sets of points $X^k = \{X_i^k = (x_{i1}^k, \ldots, x_{in}^k) \in \tilde{A},\ i = 1, \ldots, m\}, k = 1, \ldots, v$.
3. Compute v sets $S^k = \{f(X_i^k),\ i = 1, \ldots, m\}, k = 1, \ldots, v$, of values of $f(X)$ at the sets X^k.
4. Compute variances $D^* f$ on the basis of m points from the sets S^k, $k = 1, \ldots, v$. Denote the variances by $D_k^* f, k = 1, \ldots, v$.
5. Find the minimal and the maximal variance $D_{\min}^* f$ and $D_{\max}^* f$ among $D_k^* f, k = 1, \ldots, v$.
6. Divide the interval $[D_{\min}^* f,\ D_{\max}^* f]$ into p parts of equal length $\Delta = (D_{\max}^* f - D_{\min}^* f)/p$.
7. Select different $p - 2$ variances $D_{s_2}^* f, \ldots, D_{s_{p-1}}^* f$ from $D_k^* f$, $k = 1, \ldots, v$, that belong to different $p - 2$ intervals: $[D_{\min}^* f + k\Delta,\ D_{\min}^* f + (k+1)\Delta]$, $k = 1, \ldots, p - 2$.
8. The following sequence of variances will be obtained as a result of Steps 2–7: $D_{\min}^* f, D_{s_2}^* f, \ldots, D_{s_{p-1}}^* f, D_{\max}^* f$. Denote the respective numbers of sets of points and sets of values of $f(X)$ by $n_1, \ldots, n_p$.
9. Fix a pair of variables of the extremal problem (suppose x_a and x_b). Consecutively show the sequence of pictures of distributions of (η_{1a}, η_{1b}) or (η_{2a}, η_{2b}) values, calculated on the basis of data sets with the numbers defined in Step 8, in the following order:

$$n_1, n_2, \ldots, n_{p-1}, n_p, n_{p-1}, \ldots, n_2, n_1, n_2, \ldots, n_{p-1}, n_p, n_{p-1}, \ldots.$$

   In this manner, we get the effect of animation that characterizes the relationship between variables x_a and x_b.

The sequence $n_1, \ldots, n_p$ in Step 8 of the algorithm may also be based not on the variances $D_k^* f$, $k = 1, \ldots, v$, but on their logarithms: $\ln(D_k^* f)$, $k = 1, \ldots, v$. The investigator may choose this when $D^* f$ varies in a very wide range.

The investigator may also choose the method of scaling pictures presented in an animate manner:

- the distance from the closest point to the border of any picture is the same;
- the values of borders of any picture that makes up a sequence are fixed for all the pictures.

When the variance D^*f varies in a very wide range (Figure 10) the first method may be used; otherwise (Figures 11–13) – the second one.

The algorithm above may be easily combined with the algorithm of the visual analysis, presented in Section 4.2. p pictures with different distributions of η_1 or η_2 values, obtained on the basis of all possible pairs of points X_i, X_j, $i, j = 1, \dots, m$, $i \neq j$, from the set D (e.g., pictures of Figures 10–13) or via a random selection of the pairs of points from D, may be presented to the investigator in the animated manner: he will make a decision on the basis of a set of distributions that replace one another on the lapse of time.

8. Examples and Discussions

The pictures in Figure 10 are not selected by using the proposed algorithm of the animated visual analysis, because this sequence of pictures was formed in order to demonstrate a possible variety of distributions.

Figures 11–13 illustrate the performance of the algorithm via the analysis of test problems containing Branin's function f_4 from Section 3 and the function $f_5 = \exp(x_1 x_2)$, $x_1, x_2 \in [-1, 1]$. Distributions of $\eta_1(1)$ and $\eta_1(0.5)$ obtained on the basis of f_4 are presented in Figures 11 and 12, respectively, and the distributions of $\eta_1(1)$ obtained on the basis of f_5 are presented in Figure 13. $m = 30$, $v = 50$, $p = 8$ in the case of f_4, and $m = 40$, $v = 40$, $p = 8$ in the case of f_5. The second method of scaling was used. The variance D^*f is given at the left lower corner of each picture. The computer program of the algorithm is realized so that the variances $D^*_{\min}f, D^*_{s_2}f, \dots, D^*_{s_{p-1}}f, D^*_{\max}f$ are not uniformly distributed in the interval $[D^*_{\min}f, D^*_{\max}f]$. However, $D^*_{s_{k+1}}f \in [D^*_{\min}f + k\Delta, D^*_{\min}f + (k+1)\Delta]$, $k = 1, \dots, p-2$. Other, more sophisticated, realizations are possible.

Let us analyze the distributions in Figure 11 more in detail. Each picture was analyzed separately for the direction in which the density of points in the distribution of $\eta_1(1)$ is maximal, i.e. for direction $\overline{Y_1}$ in the definition domain $\tilde{A}$ that maximizes the mean absolute difference between two values of the objective function calculated by a pair randomly selected points in this direction. Two almost equivalent solutions are possible (see Section 5 and Figures 7 and 8). These two directions may be evaluated visually analyzing separate pictures of Figure 11. The directions evaluated visually are presented in circles on the left or the right of the respective pictures. Vectors that are estimates of directions $\overline{Y_1}$ separate the darker area in the circles. Such an effect with darkness allows us to easier perceive and compare the results of analysis of different pictures. The circle at the bottom of Figure 11 shows the results of visual analysis averaged through all the pictures presented to the investigator.

A similar presentation form is used in Figure 12, too. The difference is that we observe various distributions of $\eta_1(1)$ in Figure 11 and of $\eta_1(0.5)$ in Figure 12.

From the circles in Figures 11 and 12 we see that the investigator has got different estimates of directions $\overline{Y_1}$ for each picture of these figures. Which estimate is good, and which is not? It is obvious that there is no ideal estimate among them. The animated analysis of all the pictures makes it possible to get the integral decision.

Figure 14 integrates the results of visual analysis of the sets of distributions presented in Figures 11 and 12. Like in Section 5, we present here the results regarding f_4. The mean absolute difference (D1) for this function, calculated at randomly selected pairs of points in the direction whose orientation defines the angle α , is presented in Figure 7 ($\alpha \in [0, \pi]$). Dependence of D1 on α is a periodical function, the period is π. We also present the dependence of D1 on α in Figure 14, but here we use $\alpha \in [-\pi/2, \pi]$. It follows from Figure 14 that D1 has two maxima for f_4. Denote these maxima by $\max_1$ and $\max_2$. They are denoted by VA in Figure 7. Positions of the maxima of D1 are shown in the figure below the curve of D1. Distributions of the values of η_1 were investigated by the visual analysis for

1. $\tau = 1$ (see distributions analyzed in Figure 11).
2. $\tau = 0.5$ (see distributions in Figure 12).

We present the summarized results of visual analysis above the curve of D1 in Figure 14. The investigator had to analyze visually separate distributions of η_1 and to point out two directions in which the density of points is maximal. Of course, the analysis of 8 different distributions (pictures in Figure 11 or 12) yielded different decisions. The group of 8 short lines indicates these decisions. There are four such groups: estimates of $\max_1$ and $\max_2$ in two cases ($\tau = 1$ and $\tau = 0.5$). The numbers are assigned to each group in Figure 14. The arrow under the group shows the average decision obtained via the analysis

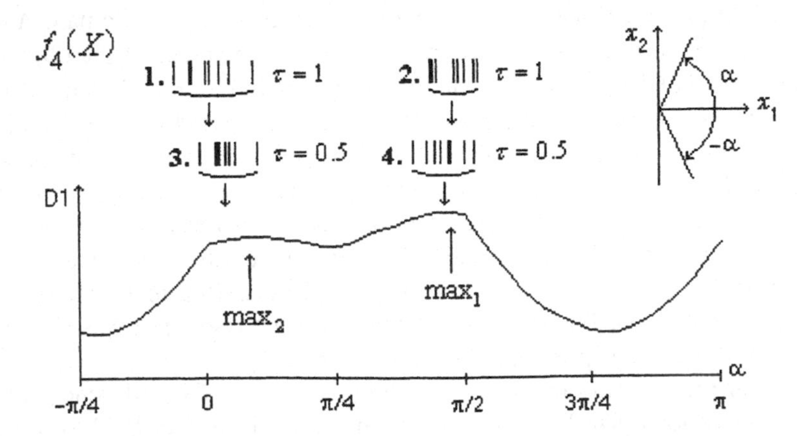

Figure 14. Results of visual analysis of the sets of distributions presented in Figures 11 and 12.

of 8 pictures (like the circle at the bottom of Figures 11 or 12 that shows the results of visual analysis averaged through all the 8 pictures presented to the investigator). We can conclude from Figure 14 that

- the variance of estimates of the direction corresponding to $\max_1$ is smaller with $\tau = 1$ compared to that for $\tau = 0.5$;
- the variance of estimates of the direction corresponding to $\max_2$ is smaller with $\tau = 0.5$ as compared to that for $\tau = 1$;
- the smaller variances of estimates are related with better averaged estimates through all the 8 pictures (see and compare a pair of groups of lines 1 and 3 or that of groups 2 and 4).

The last conclusion suggests an idea that the investigator has found some good specific and common details in the distributions of $\eta_1(1)$ (see group 2) and $\eta_1(0.5)$ (see group 3). These details influenced his decisions: the variances of obtained estimates are smaller and averaged estimates are better in these two cases.

The sequence of pictures in Figure 10 requires a special attention. We see in Figure 10 a sequence of pictures with a consequently evolving view, i.e. the view evolves from the first one to the last one, and the picture with a higher number of order includes the previous but diminished picture. Of course, the pictures appeared in random order, because the sets of points $X_i, i = 1, \ldots, m$, were generated at random. The experimental results suggest that any set of m points X_i, $i = 1, \ldots, m$, may be related with the picture of a consecutively evolving sequence. Thus, maybe we can introduce a measure of similarity of sets?

9. Conclusions

A new phenomenon that characterizes the extremal problem has been discovered. The paper tries to reveal fields of application of this phenomenon. The results presented in this paper make up the basis for a new way of analyzing extremal problems. The method of visual analysis based on the knowledge discovery in the set of objective function observations has been developed and extended to the animated analysis. The results of analysis may be used in creating a new coordinate system of the extremal problem and in a graphical representation of the observed data.

When the investigator makes a decision using the animated presentation of data sets, he can find some specific and common details in the analyzed set of distributions, better evaluate or predict the evolution of distributions, and make some approximations. The investigator has an opportunity to observe different data sets that characterize the same function for various values of τ. This also increases the quality of the final analysis.

Naturally, the presented approach to the animated visual analysis requires generating many data sets. Sometimes such a generation is very computation-expensive. Another disadvantage of the analysis is its limited application to extremal problems of high dimensionality, – the visual analysis becomes too tiresome in this case. Therefore, the ideas discussed in this paper may be applied where the investigator wants not only to solve the extremal problem, but also to discover additional knowledge on it.

Some problems remain open in this paper and make a basis for further research. They deal with

- the regularities observed visually in the analyzed distributions (these regularities indicate some specific characteristics of the structure of extremal problem; one can observe them only visually, however, what is the gain of this?);
- occurrence of a sequence of pictures with a consequently evolving view (what are the reasons? maybe, one can draw some profound conclusions basing on this?).

Bibliography

[1] Frawley, W. J., Piatetsky-Shapiro, G. and Matheus, C. J.: Knowledge discovery in databases: An overview, In: G. Piatetsky-Shapiro and C. J. Matheus (eds), *Knowledge Discovery in Databases*, AAAI Press/ The MIT Press, 1991, pp. 1–27.

[2] Fayyad, U. M., Piatetsky-Shapiro, G., Smyth, P. and Uthurusamay, R. (eds): *Advances in Knowledge Discovery and Data Mining*, AAAI Press/ The MIT Press, 1996.

[3] Dzemyda, G.: *Knowledge Discovery Seeking a Higher Optimization Efficiency*, Research Report Presented for Habilitation, Mokslo Aidai, Vilnius, 1997. ISBN 9986-479-28-2.

[4] Jones, C. V.: Visualization and optimization, *Interactive Transactions of Operations Research and Management Science (ITORMS)* **2**(1) (1998). http://orcs.bus.okstate.edu/jones98/ and http://www.chesapeake2.com/itorms/.

[5] Leipert, S., Diehl, M., Jünger, M. I. and Kupke, J.: *VBCTOOL – a Graphical Interface for Visualization of Branch Cut Algorithms the Tree Interface Version 1.0.1*, University of Cologne, 1997. http://www.informatik.uni-koeln.de/ls_juenger/projects/vbctool.html.

[6] Dean, N., Mevenkamp, M. and Monma, C. L.: Netpad: An interactive graphics system for network modeling and optimization, In: O. Balci, R. Sharda and S. A. Zenios (eds), *Computer Science and Operations Research: New Developments in Their Interfaces*, Pergamon Press, Oxford, 1992, pp. 231–243.

[7] Jones, C. V.: Animated sensitivity analysis, In: O. Balci, R. Sharda and S. A. Zenios (eds), *Computer Science and Operations Research: New Developments in Their Interfaces*, Pergamon Press, Oxford, 1992, pp. 177–196.

[8] Buchanan, I. and McKinnon, K.: An animated interactive modelling system for decision support, *European J. Oper. Res.* **54** (1991), 306–317.

[9] Carpendale, M. S. T., Cowperthwaite, D. J. and Fracchia, F. D.: 3-dimensional pliable surfaces: For the effective presentation of visual information, In: *UIST Proceedings: ACM Symposium on User Interface Software and Technology*, ACM Press, New York, 1995, pp. 217–226.

[10] Beshers, C. and Feiner, S.: AutoVisual: Rule-based design of interactive multivariate visualizations, *IEEE Comput. Graphics Appl.* **13**(4) (1993), 41–49.

[11] Chatterjee, A., Das, P. P. and Bhattacharya, S.: Visualization in linear programming using parallel coordinates, *Pattern Recognition* **26**(11) (1993), 1725–1736.

[12] Dzemyda, G.: LP-search with extremal problem structure analysis, In: N. K. Sinha and L. A. Telksnys (eds), *Proceedings of the 2nd IFAC Symposium*, IFAC Proceedings Series, No. 2, Pergamon Press, 1987, pp. 499–502.

[13] Dzemyda, G.: Visual analysis of a set of function values, In: *Proceedings of the 13th International Conference on Pattern Recognition*, Vol. 2, Track B, *Pattern Recognition and Signal Analysis*, IEEE Computer Society Press, Los Alamitos, CA 1996, pp. 700–704.

[14] Dzemyda, G.: On the visual analysis of extremal problems, *Informatica (Institute of Mathematics and Informatics, Vilnius)* **8**(2) (1997), 181–214.

[15] Šaltenis, V. and Dzemyda, G.: The structure analysis of extremal problems using some approximation of characteristics, In: A. Žilinskas (ed.), *Teorija Optimaljnych Reshenij*, Vol. 8, Inst. Math. Cybern., Vilnius, 1982, pp. 124–138 (in Russian).

[16] Šaltenis, V.: *Structure Analysis of Optimization Problems* (in Russian), Mokslas, Vilnius, 1989.

[17] Dixon, L. C. W. and Szego, G. P.: The global optimization problem: An introduction, In: L. C. W. Dixon and G. P. Szego (eds), *Towards Global Optimization 2*, North-Holland, 1978, pp. 1–15.

[18] Dzemyda, G. (ed.): *The Package of Applied Programs for Dialogue Solving of Multiextremal Problems MINIMUM: The Description of Using* (in Russian), The State Fund of Algorithms and Programs (Reg.No50860000112), Inst. Math. Cybern., Vilnius, 1985.

[19] Dzemyda, G.: Multiextremal problem of computer-aided design, *Informatica (Institute of Mathematics and Informatics, Vilnius)* **6**(3) (1995), 249–263.

[20] Reyment, R. A. and Joreskog, K. G: *Applied Factor Analysis in the Natural Sciences*, Cambridge University Press, Cambridge, 1993.

Chapter 6

TEST PROBLEMS FOR LIPSCHITZ UNIVARIATE GLOBAL OPTIMIZATION WITH MULTIEXTREMAL CONSTRAINTS

Domenico Famularo and Paolo Pugliese
DEIS, Università degli Studi della Calabria
Via Pietro Bucci 41C-42C
87036 Rende (CS), Italy
famularo;pugliese@deis.unical.it

Yaroslav D. Sergeyev
ISI-CNR, c/o DEIS
Università degli Studi della Calabria
Via Pietro Bucci 41C-42C
87036 Rende (CS), Italy
and
Software Department
University of Nizhni Novgorod
Gagarin Av. 23
Nizhni Novgorod, Russian Federation
yaro@si.deis.unical.it

Abstract In this paper, Lipschitz univariate constrained global optimization problems where both the objective function and constraints can be multiextremal are considered. Two sets of test problems are introduced, in the first one both the objective function and constraints are differentiable functions and in the second one they are non-differentiable. Each series of tests contains 3 problems with one constraint, 4 problems with 2 constraints, 3 problems with 3 constraints, and one infeasible problem with 2 constraints. All the problems are shown in figures. Lipschitz constants and global solutions are given. For each problem it is indicated whether the optimum is located on the boundary or inside a feasible subregion and the number of disjoint feasible subregions is given. Results of numerical experiments executed with the introduced test problems using Pijavskii's method combined with a non-differentiable penalty function are presented.

Keywords: Global optimization, multiextremal constraints, test problems, numerical experiments

G. Dzemyda et al. (eds.), Stochastic and Global Optimization, 93–109.
© 2002 *Kluwer Academic Publishers.*

1. Introduction

In this paper we consider the global optimization problem with nonlinear constraints

$$\min\{f(x) : x \in [a,b],\ g_j(x) \leqslant 0,\ 1 \leqslant j \leqslant m\}, \tag{1}$$

where $f(x)$ and $g_j(x), 1 \leqslant j \leqslant m$, are multiextremal Lipschitz functions (to unify the description process we shall use the designation $g_{m+1}(x) \triangleq f(x)$). More precisely, the functions $g_j(x), 1 \leqslant j \leqslant m+1$, satisfy the Lipschitz condition in the form

$$|g_j(x') - g_j(x'')| \leqslant L_j|x' - x''|, \quad x', x'' \in Q_j,\ 1 \leqslant j \leqslant m+1, \tag{2}$$

where the constants

$$0 < L_j < \infty, \quad 1 \leqslant j \leqslant m+1, \tag{3}$$

are known. Since the functions $g_j(x)$, $1 \leqslant j \leqslant m$, are supposed to be multiextremal, the subdomains $Q_j \subset [a,b]$, $2 \leqslant j \leqslant m+1$, can have a few disjoint subregions each. In the following we shall suppose that all the sets Q_j, $2 \leqslant j \leqslant m+1$, either are empty or consist of a finite number of disjoint intervals of a finite positive length.

The recent literature in constrained optimization [1–12] practically does not contain sets of tests with multiextremal constraints. This paper introduces problems for a systematic comparison of numerical algorithms developed for solving the global optimization problems with multiextremal constraints. Performance of the method of Pijavskii [4,6,13] combined with a non-differentiable penalty function on the introduced test problems is shown.

Two series of problems (ten feasible and one infeasible problem each) have been developed. The first series of tests is based on problems where both the objective function and the constraints are differentiable. The second series consists of problems where both the objective function and constraints are non-differentiable. Each series of tests contains:

- 3 problems with one constraint;
- 4 problems with 2 constraints;
- 3 problems with 3 constraints;
- one infeasible problem with 2 constraints.

For each problem the number of disjoint feasible subregions is presented. It is indicated whether the optimum is located on the boundary or inside a feasible subregion. All the problems are shown in Figures 1 and 2 (Differentiable

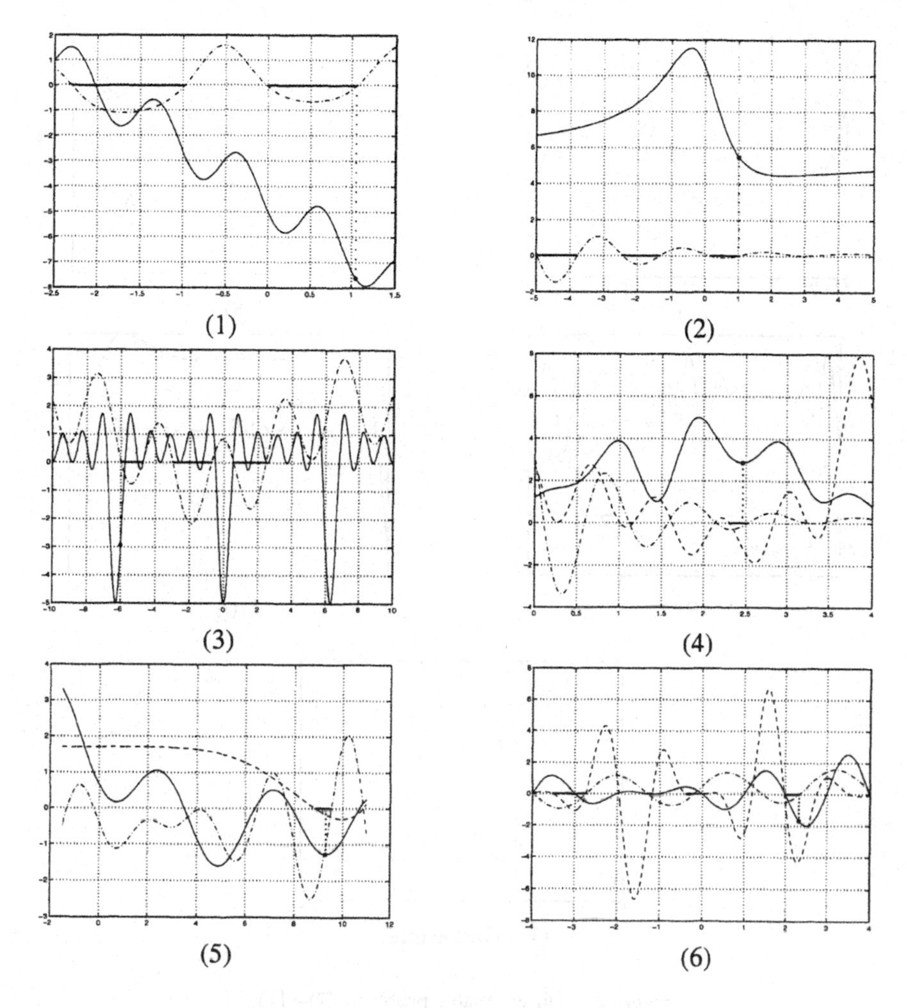

Figure 1. Differentiable problems (1)–(6).

problems) and Figures 3 and 4 (Non-differentiable problems). The constraints are drawn by dotted/mix-dotted lines and the objective function is drawn by a solid line. The feasible region is described by a collection of bold segments on the x axis and the global solution is represented by an asterisk located on the graph of the objective function.

In Table 1 (Differentiable problems) and Table 2 (Non-differentiable problems) the Lipschitz constants of the objective function and the constraints are reported together with an approximation of the global solution $(x^*, f(x^*))$. All these quantities have been computed by sweeping iteratively the search interval $[a, b]$ by the step $10^{-6}(b-a)$. Note that all the constrained problems have a unique global solution.

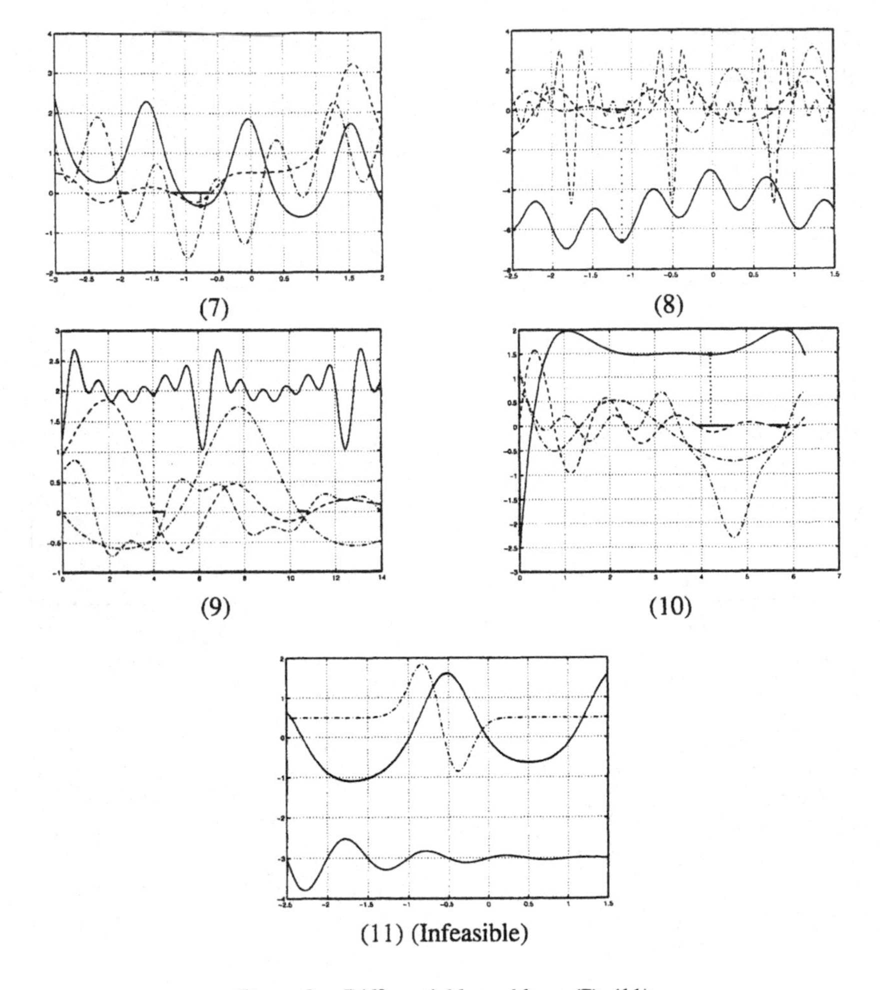

Figure 2. Differentiable problems (7)–(11).

2. Differentiable Problems

In this section the problems where both the objective function and the constraints are differentiable are described (see Table 1).

PROBLEM 1.

$$\min_{x \in [-2.5, 1.5]} f(x) = -\frac{13}{6}x + \sin\left(\frac{13}{4}(2x+5)\right) - \frac{53}{12}$$

subject to

$$g_1(x) = \exp(-\sin(3x)) - \frac{1}{10}\left(x - \frac{1}{2}\right)^2 - 1 \leqslant 0.$$

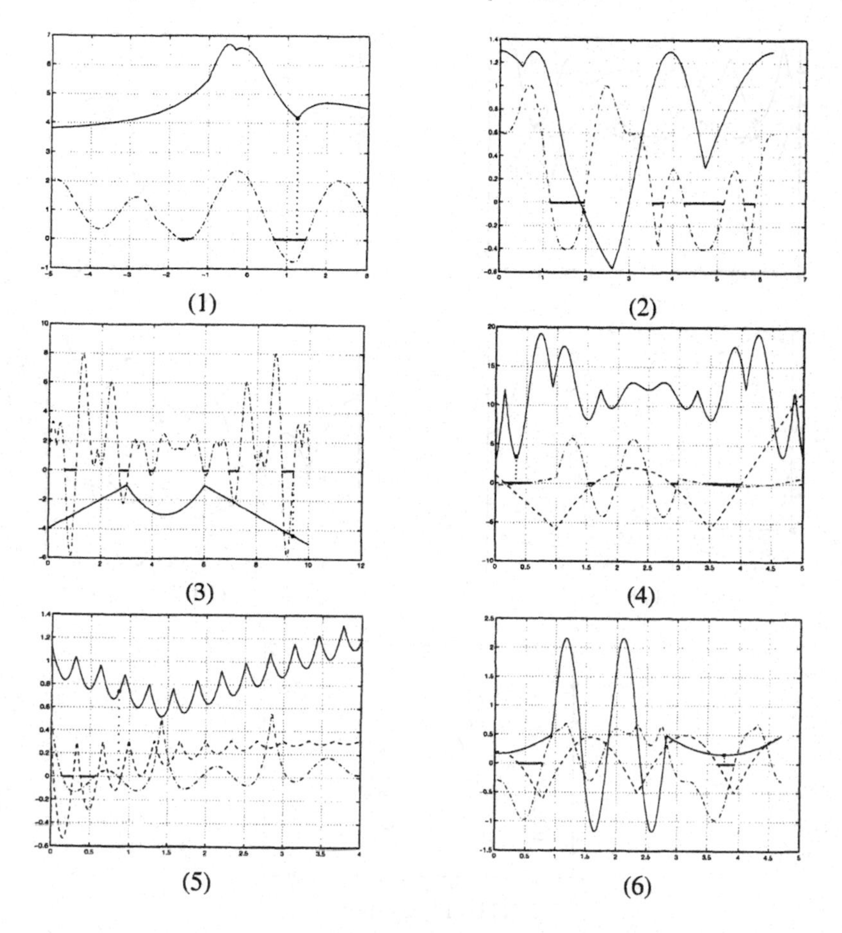

Figure 3. Non-differentiable problems (1)–(6).

The problem has 2 disjoint feasible subregions and the global optimum x^* is located on the boundary of one of the feasible subregions (see Figure 1(1)).

PROBLEM 2.

$$\min_{x\in[-5,5]} 2f(x) = \frac{11x^2 - 10x + 21}{2(x^2+1)}$$

subject to

$$g_1(x) = \frac{1}{20} - \exp\left(-\frac{2}{5}(x+5)\right)\sin\left(\frac{4}{5}\pi(x+5)\right) \leqslant 0.$$

The problem has 3 disjoint feasible subregions and the global optimum x^* is located on the boundary of one of the feasible subregions (see Figure 1(2)).

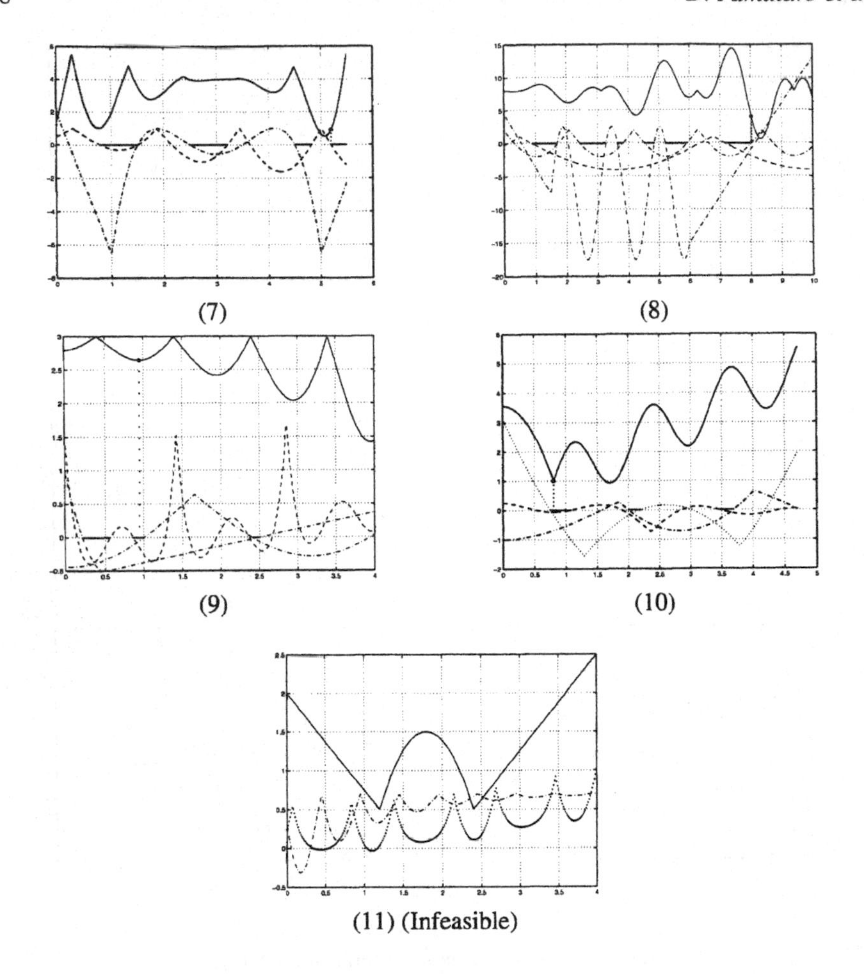

Figure 4. Non-differentiable problems (7)–(11).

PROBLEM 3.

$$\min_{x \in [-10,10]} f(x) = -\sum_{i=1}^{5} \cos(ix)$$

subject to

$$g_1(x) = \frac{3}{2}\left(\cos\left(\frac{7}{20}(x+10)\right) - \sin\left(\frac{7}{4}(x+10)\right) + \frac{1}{2}\right) \leqslant 0.$$

The problem has 3 disjoint feasible subregions and the global optimum x^* is located on the boundary of one of the feasible subregions (see Figure 1(3)).

Table 1. Differentiable problems. Lipschitz constants and global solutions

	Lipschitz Constants				Global Solutions	
Pr.	$g_1(x)$	$g_2(x)$	$g_3(x)$	$f(x)$	x^*	$f(x^*)$
1	4.640837	–	–	8.666667	1.05738	−7.61284448
2	2.513269	–	–	6.372595	1.016	5.46063488
3	3.124504	–	–	13.201241	−5.9921	−2.94600839
4	29.731102	35.390605	–	12.893183	2.45956	2.8408089
5	5.654617	0.931981	–	2.021595	8.85725	−1.27299809
6	2.480000	25.108154	–	8.835339	2.32396	−1.6851399
7	8.332010	5.359309	–	6.387862	−0.774575	−0.33007413
8	20.184982	90.598898	6.372137	10.415012	−1.12724	−6.60059665
9	0.873861	1.682731	1.254588	3.843648	4.0000	1.92218867
10	3.170468	4.329008	7.999984	12.442132	4.2250023	1.474
11	4.640837	10.000000	–	6.283173	–	–

Table 2. Non-differentiable problems. Lipschitz constants and global solutions

	Lipschitz Constants				Global Solutions	
Pr.	$g_1(x)$	$g_2(x)$	$g_3(x)$	$f(x)$	x^*	$f(x^*)$
1	3.808540	–	–	3.499998	1.25832	4.17418934
2	3.404631	–	–	2.000000	1.95966267	−0.07913964
3	47.250828	–	–	2.666662	9.40115	−4.40115
4	31.415927	12.799987	–	75.819889	0.33295	3.3461957
5	5.557050	9.424759	–	2.750000	0.86992	0.74162802
6	4.577345	2.166549	–	11.111111	3.76991118	0.16666667
7	21.999866	5.436564	–	23.400533	5.20115750	0.90278234
8	40.000000	6.000000	2.500000	23.625414	8.0285	4.0470244
9	1.050000	5.999988	16.671149	4.007294	0.95024	2.64804101
10	1.887454	2.334828	4.949998	6.399980	0.79999872	1.00000822
11	5.205608	6.921230	–	3.333328	–	–

PROBLEM 4.

$$\min_{x\in[0,4]} f(x) = 4\sin\left(\frac{\pi}{4}x+\frac{1}{20}\right)\left(\sin^3\left(\frac{\pi}{2}x+\frac{1}{10}\right)+\cos^3\left(\frac{\pi}{2}x+\frac{1}{10}\right)\right)^2+1$$

subject to

$$g_1(x) = \frac{6}{25} - 2\sum_{i=1}^{5}\cos\left(\frac{5}{4}(i+1)x+i\right) \leqslant 0,$$

$$g_2(x) = \frac{9}{50} - \frac{9}{2}\exp\left(-\left(x-\frac{1}{10}\right)\right)\sin\left(2\pi\left(x-\frac{1}{10}\right)\right) \leqslant 0.$$

The problem has 2 disjoint feasible subregions and the global optimum x^* is located inside one of the feasible subregions (see Figure 1(4)).

PROBLEM 5.

$$\begin{aligned}\min_{x\in[-1.5,11]} f(x) = {} & \sin(0.423531x + 3.13531) + \\ & + \sin\left(\frac{10}{3}(0.423531x + 3.13531)\right) + \\ & + \log(0.423531x + 3.13531) + 0.36634 - 0.355766x\end{aligned}$$

subject to

$$\begin{aligned}g_1(x) = {} & \frac{17}{25} - \frac{2}{29763.233} \times \\ & \times\left(-\frac{1}{6}x^6 + \frac{52}{25}x^5 - \frac{39}{80} - \frac{71}{10}x^3 + \frac{79}{20}x^2 + x - \frac{1}{10}\right) \leqslant 0,\\ g_2(x) = {} & -\frac{14}{125}(3x - 8)\sin\left(\frac{252}{125}\left(x + \frac{3}{2}\right)\right) - \frac{1}{2} \leqslant 0.\end{aligned}$$

The problem has one disjoint feasible region and the global optimum x^* is located inside the feasible region (see Figure 1(5)).

PROBLEM 6.

$$\min_{x\in[-4,4]} f(x) = -\frac{7}{40}(3x + 4)\sin\left(\frac{63}{20}(x + 4)\right)$$

subject to

$$g_1(x) = 40\left(\cos(4x)(x - \sin(x))\exp(-\frac{x^2}{2})\right) \leqslant 0,$$

$$g_2(x) = \frac{2}{25}(x + 4) - \sin\left(\frac{12}{5}(x + 4)\right) \leqslant 0.$$

The problem has 4 disjoint feasible subregions and the global optimum x^* is located on the boundary of one of the feasible subregions (see Figure 1(6)).

PROBLEM 7.

$$\min_{x\in[-3,2]} f(x) = \exp(-\cos(4x - 3)) + \frac{1}{250}(4x - 3)^2 - 1$$

subject to

$$g_1(x) = \sin^3(x)\exp(-\sin(3x)) + \frac{1}{2} \leqslant 0,$$

$$g_2(x) = \cos\left(\frac{7}{5}(x + 3)\right) - \sin(7(x + 3)) + \frac{3}{10} \leqslant 0.$$

The problem has 2 disjoint feasible subregions and the global optimum x^* is located inside one of the feasible subregions (see Figure 2(7)).

PROBLEM 8.

$$\min_{x\in[-2.5,1.5]} f(x) = \cos\left(\frac{7}{4}x + \frac{241}{40}\right) - \sin\left(\frac{35}{4}x + \frac{241}{8}\right) - 5$$

subject to

$$g_1(x) = \exp(-\sin(4x)) - \frac{1}{10}(x - \frac{1}{2})^2 - 1 \leqslant 0,$$

$$g_2(x) = \frac{3}{10} - \sum_{i=1}^{5} \cos\left(5(i+1)\left(x + \frac{1}{2}\right)\right) \leqslant 0,$$

$$g_3(x) = \left(-\frac{21}{20}x - \frac{13}{8}\right)\sin\left(\frac{63}{10}x + \frac{63}{4}\right) + \frac{1}{5} \leqslant 0.$$

The problem has 3 disjoint feasible subregions and the global optimum x^* is located inside one of the feasible subregions (see Figure 2(8)).

PROBLEM 9.

$$\min_{x\in[0,14]} f(x) = \sum_{i=1}^{5} \frac{1}{5}\sin((i+1)x - 1) + 2$$

subject to

$$g_1(x) = \frac{1}{40}(x-4)\left(x - \frac{32}{5}\right)(x-9)(x-11)\exp\left(-\frac{1}{10}\left(x - \frac{13}{2}\right)^2\right) \leqslant 0,$$

$$g_2(x) = (\sin^3(x+1) + \cos^3(x+1))\exp\left(-\frac{x+1}{10}\right) \leqslant 0,$$

$$g_3(x) = \exp\left(-\cos\left(\frac{3}{5}\left(x - \frac{5}{2}\right)\right)\right) + \frac{1}{10}\left(\frac{3}{25}x - \frac{4}{5}\right)^2 - 1 \leqslant 0.$$

The problem has 2 disjoint feasible subregions and the global optimum x^* is located on the boundary of one of the feasible subregions (see Figure 2(9)).

PROBLEM 10.

$$\min_{x\in[0,2\pi]} f(x) = -\frac{1}{500}\left(\frac{4}{\pi}\left(x - \frac{3}{10}\right) - 4\right)^6 + \frac{3}{100}\left(\frac{4}{\pi}\left(x - \frac{3}{10}\right) - 4\right)^4 - \frac{27}{500}\left(\frac{4}{\pi}\left(x - \frac{3}{10}\right) - 4\right)^2 + \frac{3}{2}$$

subject to

$$g_1(x) = 2\exp\left(-\frac{2}{\pi}x\right)\sin(4x) \leqslant 0,$$

$$g_2(x) = -\left(\frac{2}{\pi}x - \frac{1}{2}\right)^2 \frac{(-(\frac{2}{\pi}x - \frac{1}{2})^2 + 5(\frac{2}{\pi}x - \frac{1}{2}) - 6)}{((\frac{2}{\pi}x - \frac{1}{2})^2 + 1)} - \frac{1}{2} \leqslant 0,$$

$$g_3(x) = \sin^3(x) + \cos^3(2x) - \frac{3}{10} \leqslant 0.$$

The problem has 2 disjoint feasible subregions and the global optimum x^* is located inside one of the feasible subregions (see Figure 2(10)).

PROBLEM 11.

$$\min_{x\in[-2.5,1.5]} f(x) = -\exp\left(-\left(x + \frac{5}{2}\right)\right)\sin\left(2\pi\left(x + \frac{5}{2}\right)\right) - 3$$

subject to

$$g_1(x) = \exp(-\sin(3 * x)) - \frac{1}{10}\left(x - \frac{1}{2}\right)^2 - 1 \leqslant 0,$$

$$g_2(x) = \frac{1}{2} - \left(10\exp\left(-10\left(x + \frac{3}{5}\right)\right)\sin\left(x + \frac{3}{5}\right)\right) \leqslant 0.$$

The problem is infeasible (see Figure 2(11)).

3. Non-differentiable Problems

In this section the problems where both the objective function and the constraints are non-differentiable are described in Table 2 and Figures 3 and 4.

PROBLEM 1.

$$\min_{x\in[-5,3]} f(x) = \left|\frac{x^2 - 10x + 11}{2(x^2 + 1)}\right| + \left|\frac{3x^2 + 4x + 1}{x^2 + 1}\right|$$

subject to

$$g_1(x) = \left|\sin\left(\frac{7}{554}(69x + 347)\right) + \cos\left(\frac{7}{554}(69x + 347)\right)\right| + \\ + \cos\left(\frac{21}{554}(69x + 347)\right) \leqslant 0.$$

The problem has 2 disjoint feasible subregions and the global optimum x^* is located inside one of the feasible subregions (see Figure 3(1)). Note also that the minimum is located on a point at which the derivative is not continuous.

PROBLEM 2.

$$\min_{x\in[0,2\pi]} f(x) = \max\{\sin(2x), \cos(x)\} + \frac{3}{10}$$

subject to

$$g_1(x) = |\sin^3(2x) + \cos^3(x)| - \frac{2}{5} \leqslant 0.$$

The problem has 4 disjoint feasible subregions and the global optimum x^* is located on the boundary of one of the feasible subregions (see Figure 3(2)).

PROBLEM 3.

$$\min_{x\in[0,10]} f(x) = \begin{cases} x-4, & x \leqslant 3, \\ \frac{8}{9}x^2 - 8x + 15, & 3 < x \leqslant 6, \\ -x+5, & x > 6 \end{cases}$$

subject to

$$g_1(x) = \frac{3}{2} - \cos(6(x-5))|2(x-5)\sin(2(x-5))| \leqslant 0.$$

The problem has 6 disjoint feasible subregions and the global optimum x^* is located on the boundary of one of the feasible subregions (see Figure 3(3)).

PROBLEM 4.

$$\min_{x\in[0,5]} f(x) = 2\cos(4x-10)|(4x-10)\sin(4x-10)| + 12$$

subject to $g_1(x) \leqslant 0$, $g_2(x) \leqslant 0$ where

$$g_1(x) = \begin{cases} -7x+1, & x \leqslant 1, \\ -\frac{128}{25}x^2 + \frac{576}{25}x - \frac{598}{25}, & 1 < x \leqslant \frac{7}{2}, \\ 12(x-4), & x > \frac{7}{2}, \end{cases}$$

$$g_2(x) = \begin{cases} x^2 - \frac{3}{10}, & x \leqslant 1, \\ 5\sin(2\pi x) + \frac{7}{10}, & 1 < x \leqslant 3, \\ x^2 - 8x + \frac{163}{10}, & x > 3. \end{cases}$$

The problem has 4 disjoint feasible subregions and the global optimum x^* is located inside one of the feasible subregions (see Figure 3(4)).

PROBLEM 5.

$$\min_{x\in[0,4]} f(x) = \frac{1}{4}\left(\left|x - \frac{3}{2}\right| - |\sin(10x)| + 3\right)$$

subject to

$$g_1(x) = \frac{8}{25} - \exp(-x)|\sin(3\pi x)| \leqslant 0,$$
$$g_2(x) = \exp\left(-\left|\sin\left(\frac{5}{2}\sin\left(\frac{11}{5}x\right)\right)\right|\right) - \frac{1}{2} + \frac{1}{100}x^2 \leqslant 0.$$

The problem has 3 disjoint feasible subregions and the global optimum x^* is located on the boundary of one of the feasible subregions (see Figure 3(5)).

PROBLEM 6.

$$\min_{x\in[0,1.5\pi]} f(x) = \begin{cases} \frac{1}{3}\left(\frac{100}{9\pi^2}x^2 + \frac{1}{2}\right), & x \leqslant \frac{3\pi}{10}, \\ \frac{5}{3}\sin\left(\frac{20}{3}x\right) + \frac{3}{2}, & \frac{3\pi}{10} < x \leqslant \frac{9\pi}{10}, \\ \frac{1}{3}\left(\frac{100}{9\pi^2}x^2 - \frac{80}{3\pi}x + \frac{33}{2}\right), & x > \frac{9\pi}{10} \end{cases}$$

subject to

$$g_1(x) = -\left|\frac{(x-\pi)^3}{100}\right| + |\cos(2(x-\pi))| - \frac{1}{2} \leqslant 0,$$

$$g_2(x) = \frac{7}{10} - |\sin^3(3x) + \cos^3(x)| \leqslant 0.$$

The problem has 2 disjoint feasible subregions and the global optimum x^* is located inside one of the feasible subregions (see Figure 3(6)).

PROBLEM 7.

$$\min_{x\in[0,5.5]} f(x) = 4 - \frac{4}{3}\left(x - \frac{31}{10}\right)^2 \sin\left(\frac{1}{4}\left(x + \frac{9}{5}\right)\right)\left(\left|\sin\left(3x + \frac{27}{5}\right)\right| - \frac{3}{10}\right)$$

subject to $g_1(x) \leqslant 0$, $g_2(x) \leqslant 0$ where

$$g_1(x) = -\left|\sin\left(2\left(x - \frac{3}{10}\right)\right)\right| \exp\left(\sin\left(\frac{1}{3}\left(x - \frac{3}{10}\right)\right)\right) + 1,$$

$$g_2(x) = \begin{cases} -\frac{137}{16}x + 2, & x \leqslant 1, \\ -\left(x - \frac{3}{2}\right)\left(x - \frac{5}{2}\right)\left(x - \frac{7}{2}\right)\left(x - \frac{9}{2}\right), & 1 < x \leqslant 5, \\ \frac{137}{16}x - \frac{395}{8}, & x > 5. \end{cases}$$

The problem has 4 disjoint feasible subregions and the global optimum x^* is located on the boundary of one of the feasible subregions (see Figure 4(7)).

PROBLEM 8.

$$\min_{x \in [0,10]} f(x) = -\cos(3x)|x \sin(x)| + 8$$

subject to

$$g_1(x) = -5\left|\sin\left(\frac{1}{2}\left(x - \frac{1}{2}\right)\right)\right| + 1 \leqslant 0,$$

$$g_2(x) = -4\left|\sin\left(\frac{3}{2}x\right)\right| + 2 \leqslant 0,$$

$$g_3(x) = \begin{cases} -8x + \frac{9}{2}, & x \leqslant \frac{3}{2}, \\ 10\sin\left(4\left(x - \frac{3}{2}\right)\right) - \frac{15}{2}, & \frac{3}{2} < x \leqslant 6 \leqslant 0, \\ 7x - \frac{99}{2} + 10\sin(18), & x > 6. \end{cases}$$

The problem has 6 disjoint feasible subregions and the global optimum x^* is located on the boundary of one of the feasible subregions (see Figure 4(8)).

PROBLEM 9.

$$\min_{x \in [0,4]} f(x) = 3 - 2\exp\left(-\frac{1}{2}\left(\frac{22}{5} - x\right)\right)\left|\sin\left(\pi\left(\frac{22}{5} - x\right)\right)\right|$$

subject to

$$g_1(x) = 3\left(\exp\left(-\left|\sin\left(\frac{5}{2}\sin\left(\frac{11}{5}x\right)\right)\right|\right) + \frac{1}{100}x^2 - \frac{1}{2}\right) \leqslant 0,$$

$$g_2(x) = \begin{cases} 6\left(x - \frac{1}{2}\right)^2 - \frac{1}{2}, & x \leqslant \frac{1}{2} \\ \frac{1}{4}\left(x - \frac{5}{2}\right), & x > \frac{1}{2} \end{cases} \leqslant 0,$$

$$g_3(x) = \frac{4}{5} - \left(\left|\sin\left(\frac{24}{5} - x\right)\right| + \frac{6}{25} - \frac{x}{20}\right) \leqslant 0.$$

The problem has 3 disjoint feasible subregions and the global optimum x^* is located inside one of the feasible subregions (see Figure 4(9)).

PROBLEM 10.

$$\min_{x\in[0,1.5\pi]} f(x) = \begin{cases} -4x^2 + \frac{89}{25}, & x \leqslant \frac{4}{5}, \\ \sin(5x - 4) + x - \frac{1}{5}, & x > \frac{4}{5} \end{cases}$$

subject to

$$g_1(x) = \max\left\{\left(x - \frac{3}{4}\right)\left(x - \frac{21}{5}\right), -\left(x - \frac{11}{5}\right)\left(x - 3\right)\right\} \leqslant 0,$$

$$g_2(x) = -\max\left\{-\left(x - \frac{37}{10}\right)(x - 2), \cos(x)\right\} \leqslant 0,$$

$$g_3(x) = \exp\left(-\frac{x}{20}\right)\left|\sin^3(x) + \cos^3(x)\right| - \frac{3}{4} \leqslant 0.$$

The problem has 3 disjoint feasible subregions and the global optimum x^* is located inside one of the feasible subregions (see Figure 4(10)).

PROBLEM 11.

$$\min_{x\in[0,4]} f(x) = \begin{cases} \frac{1}{2}\left(4 - \frac{5}{2}x\right), & x \leqslant \frac{6}{5}, \\ \frac{1}{2}\left(-\frac{50}{9}x^2 + 20x - 15\right), & \frac{6}{5} < x \leqslant \frac{12}{5}, \\ \frac{5}{2}\left(\frac{x}{2} - 1\right), & x > \frac{12}{5} \end{cases}$$

subject to

$$g_1(x) = \exp\left(-\left|\cos\left(\frac{13}{5}\sin\left(\frac{12}{5}\left(x+\frac{1}{5}\right)\right)\right)\right|\right) - \frac{9}{20} + \frac{1}{36}\left(x+\frac{1}{5}\right)^2 \leqslant 0,$$

$$g_2(x) = \frac{7}{10} - \exp\left(-\left(x-\frac{1}{5}\right)\right)\left|\cos\left(2\pi\left(x-\frac{1}{5}\right)\right)\right| \leqslant 0.$$

The problem is infeasible (see Figure 4(11)).

4. Numerical Experiments

In this section the method proposed by Pijavskii (see [5,6,13]) has been tested on the problems described in the previous Sections. Since the method [13] works with problems having box constraints, in the executed experiments the constrained problems were reduced by the method of penalty functions to such a form. The same accuracy $\varepsilon = 10^{-4}(b-a)$ (where b and a represent the extrema of the optimization interval) has been used in all the experiments.

In Table 3 (Differentiable problems) and Table 4 (Non-differentiable problems) the results obtained by the method of Pijavskii are collected. The constrained problems were reduced to the unconstrained ones as follows

$$f_{P^*}(x) = f(x) + P^* \max\{g_1(x), g_2(x), \ldots, g_{N_v}(x), 0\}. \tag{4}$$

The coefficient P^* has been computed by the rules:

1. the coefficient P^* has been chosen equal to 15 for all the problems and it has been checked if the found solution (XPEN, FXPEN) for each problem belongs or not to the feasible subregions;
2. if it does not belong to the feasible subregions, the coefficient P^* has been iteratively increased by 10 starting from 20 until a feasible solution has been found. Particularly, this means that a feasible solution has not been found in Table 3 for the Problem 2 when P^* is equal to 80, for the Problem 4 when P^* is equal to 480, and in Table 4 for the Problem 5 when P^* is equal to 15.

It must be noticed that in Tables 3, 4 the column "Evaluations" shows the total number of evaluations of the objective function $f(x)$ and all the constraints. Thus, it is equal to

$$(N_v + 1) \times N_{\text{iter}},$$

where N_v is the number of constraints and N_{iter} is the number of iterations for each problem.

5. A Brief Conclusion

In this paper, 22 test problems for Lipschitz univariate constrained global optimization have been proposed. All the problems have both the objective

Table 3. Differentiable functions. Numerical results obtained by the method of Pijavskii working with the penalty function (4)

Problem	XPEN	FXPEN	P^*	Iterations	Eval.
1	1.05718004	−7.61185807	15	83	166
2	1.01609254	5.46142698	90	954	1906
3	−5.99184997	−2.94292577	15	119	238
4	2.45953057	2.84080890	490	1762	5286
5	9.28468704	−1.27484673	15	765	2295
6	2.32334492	−1.68307049	15	477	1431
7	−0.77476915	−0.33007412	15	917	2751
8	−1.12719146	−6.60059658	15	821	3284
9	4.00042801	1.92220821	15	262	1048
10	4.22482084	1.47400000	15	2019	8076
Average	–	–	–	817.9	2648.1

Table 4. Non-differentiable problems. Numerical results obtained by the method of Pijavskii working with the penalty function (4)

Problem	XPEN	FXPEN	P^*	Iterations	Eval.
1	1.25810384	4.17441502	15	247	494
2	1.95953624	−0.07902265	15	241	482
3	9.40072023	−4.40072023	15	797	1594
4	0.33278550	3.34620350	15	272	819
5	0.86995489	0.74168456	20	671	2013
6	3.76944805	0.16666667	15	909	2727
7	5.20113260	0.90351752	15	199	597
8	8.02859874	4.05157770	15	365	1460
9	0.95019236	2.64804101	15	1183	4732
10	0.79988668	1.00072517	15	135	540
Average	–	–	–	501.9	1545.8

function and constraints multiextremal. The problems have been collected in two sets. The first one contains tests with both the objective function and constraints being differentiable functions. The second set of tests contains problems with non-differentiable functions. Each series of tests consists of 3 problems with one constraint, 4 problems with 2 constraints, 3 problems with 3 constraints, and one infeasible problem with 2 constraints.

All the test problems have been studied in depth. Each problem has been provided with:

- an accurate estimate of the global solution;

- an accurate estimate of Lipschitz constants for the objective functions and constraints;

- figure showing the problem with the global solution and the feasible region with indication of the number of disjoint feasible subregions;
- indication whether the optimum was located on the boundary or inside a feasible subregion.

Numerical experiments with the introduced test problems have been executed with Pijavskii's method combined with a non-differentiable penalty function.

Bibliography

[1] Bomze, I. M., Csendes, T., Horst, R. and Pardalos, P. M.: *Developments in Global Optimization*, Kluwer Acad. Publ., Dordrecht, 1997.

[2] Floudas, C. A., Pardalos, P. M., Adjiman, C., Esposito, W. R., Gümüs, Z. H., Harding, S. T., Klepeis, J. L., Meyer, C. A. and Schweiger, C. A.: *Handbook of Test Problems in Local and Global Optimization*, Kluwer Acad. Publ., Dordrecht, 1999.

[3] Floudas, C.A. and Pardalos, P. M.: *State of the Art in Global Optimization*, Kluwer Acad. Publ., Dordrecht, 1996.

[4] Hansen, P., Jaumard, B. and Lu, S.-H.: Global optimization of univariate Lipschitz functions: 1. Survey and properties, *Math. Programming* **55** (1992), 251–272.

[5] Hansen P., Jaumard, B. and Lu, S.-H.: Global optimization of univariate Lipschitz functions: 2. New algorithms and computational comparison, *Math. Programming* **55** (1992), 273–293.

[6] Horst R. and Pardalos, P. M.: *Handbook of Global Optimization*, Kluwer Acad. Publ., Dordrecht, 1995.

[7] Horst R. and Tuy, H.: *Global Optimization – Deterministic Approaches*, Springer-Verlag, Berlin, 1993.

[8] Mockus, J.: *Bayesian Approach to Global Optimization*, Kluwer Acad. Publ., Dordrecht, 1988.

[9] Mockus, J., Eddy, W., Mockus, A., Mockus, L. and Reklaitis, G.: *Bayesian Heuristic Approach to Discrete and Global Optimization: Algorithms, Visualization, Software, and Applications*, Kluwer Acad. Publ., Dordrecht, 1996.

[10] Nocedal, J. and Wright, S. J.: *Numerical Optimization*, Springer Ser. Oper. Res., Springer-Verlag, 1999.

[11] Strongin, R. G. and Sergeyev, Ya. D.: *Global Optimization with Non-Convex Constraints: Sequential and Parallel Algorithms*, Kluwer Acad. Publ., Dordrecht, 2000.

[12] Sun, X. L. and Li, D.: Value-estimation function method for constrained global optimization, *J. Optim. Theory Appl.* **102**(2) (1999), 385–409.

[13] Pijavskii, S. A.: An algorithm for finding the absolute extremum of a function, *USSR Comput. Math. Math. Phys.* **12** (1972), 57–67.

Chapter 7

NUMERICAL TECHNIQUES IN APPLIED MULTISTAGE STOCHASTIC PROGRAMMING

Karl Frauendorfer and Gido Haarbrücker

Institute for Operations Research

University of St. Gallen

Switzerland

Abstract This contribution deals with the apparent difficulties when solving an optimization problem with random influences by the use of multistage stochastic linear programming. It names specific numerical solution techniques which are suitable for coping with the curse of dimensionality, with increasing scenario trees and associated large-scale LPs. The focus lies on classic decomposition methods which are natural candidates to apply parallelization techniques. In addition, an alternative approach is sketched where the replacement of optimization runs by optimality checks leads to an efficient handling of consecutive discretization steps given some structural requirements are fulfilled.

Keywords: Multistage stochastic linear programming, discretization, large-scale linear program, numerical techniques

1. Introduction

A huge number of operational and planning problems is characterized by sequences of decisions over time. As though this structure of acting over time – in the sense of taking optimal or, at least, "good" decisions w.r.t. some constraints – would not be complex enough, one often is additionally faced with a stochastic evolution of parts of the problem's parameters: prices, sales, production costs or interest rates are only a few examples of relevant future information which is very rarely already known today. Whenever possible, the today's and future decisions should take into account this uncertainty and respond to the realizations of the random variables involved.

Then, the resulting model for optimal decision making is a *multistage stochastic program* given the following assumptions are fulfilled:

G. Dzemyda et al. (eds.), Stochastic and Global Optimization, 111–127.
© 2002 *Kluwer Academic Publishers.*

- These distributions are unaffected by the decisions taken.
- Decisions have to be taken on past information only, i.e. decisions made today cannot depend on information received tomorrow or any day thereafter (so-called *nonanticipativity*).
- A finite number of decision stages is considered.

As to the third item, finitely many stages do not demand for a finite horizon of the problem to be formulated; for instance, the problem may in fact be characterized by an infinite planning horizon, but this circumstance has to be incorporated in the model by a finite number of stages.

It is possible that in some – perhaps most practical – situations a distribution law of the stochastic influences may not be known. In such cases, one can choose an approximating distribution to achieve results which normally are superior to a substitution of the random parameters by single best estimates or mean values. Anyway, at the end, one needs a concrete probability distribution (an a priori known one, a subjective idea, an approximating one or whatsoever) because otherwise one could just formulate the stochastic program without any chance to determine a solution. For this reason, in the following we will assume that the probability distributions of stochastic influences in all stages are known.

In general, it is the expected value of an interesting quantity which will be maximized or minimized subject to some constraints. Throughout this paper, we confine ourselves to multistage stochastic programs with fixed recourse for two reasons: first, these programs proved to be an adequate tool reflecting the complexity of real-life situations and, secondly, in the last decades a wide theory and powerful solution procedures have been developed. Furthermore, the objective function is assumed to be linear in the decision variables so that we are dealing with multistage stochastic *linear* programs (MSLP) with fixed recourse. For ease of simplicity, we take the *technology matrices* T^t to be known as well, but they can differ from stage to stage denoted by the superscript t.

When stochastic influence is given by continuous distributions, one of the most common – and in fact promising – approaches is to discretize the support of these distributions. Of course, although we concentrate on discretization approaches, we do not wish to conceal the existence of alternative methods: for instance, approximation of the original distributions can also be achieved by simpler – but still continuous! – distributions (see, e.g., Birge and Wallace [1], Birge and Wets [2], Wallace [3], and Birge and Qi [4]). However, discretization always leads to scenario trees depicting the stochastic evolution. On the one hand, this overcomes the obstacle of numerical integration in high dimensions corresponding to the random variables because an MSLP with a finite number of scenarios still has a *deterministic equivalent linear program.* On the other hand, such a MSLP in its extensive form mostly is given by a really large-scale linear program; normally, the latter does not appear accessible to manipulations such as *basis factorizations* for extreme or interior point methods (see, e.g., Birge and Louveaux [5], Chapter 5, for factorization in the two-stage case). Up to a certain size, these large-scale LPs are solvable by *direct methods* like simplex-based or interior point methods with acceptable computational effort. When the size of an LP sets limits to the application of direct methods, promising *indirect* methods are given by methods which are based on *decompositions*, on some form of *Lagrangian relaxation*, and on uses of *separability*.

Anyway, the solutions based on scenarios and, hence, on approximating discrete probability measures are solutions of approximate programs; they can only serve as approximate solutions of the original 'continuously distributed' MSLP. Thus, one normally aims at a successive improvement of the approximate solution until an a priori approximation quality[1] is reached. This leads to a cycle of consecutive approximation steps ν which are – hopefully – tractable until the approximation quality desired is achieved.

To summarize, two aspects are crucial to "solve" an MSLP with continuous underlying distributions.

(i) The discretization scheme is responsible for the quality of the scenarios, and it determines the type of result one can get based on these scenarios. Does one achieve probabilistic or deterministic bounds on the true[2] expected objective value? And are the scenarios representative for the stochastic evolution of the uncertain components (in case of probabilistic bounds), or are they adequately determined to actually provide bounds, respectively?

(ii) Depending on methodological aspects of the approach utilized and/or the discretization scheme, one is forced to dispose of a more or less powerful (direct or indirect) solution method. In general, it holds: the higher the number of scenarios to be dealt with, or the better the demanded approximation quality, or the larger the single LPs to be solved in approximation step ν, the sooner one gets into trouble with one's solution method.

There are many ways to approach this two-sided medal: for instance, (1) use a large crude scenario sample in conjunction with a sophisticated – e.g., parallelized – method to solve the corresponding large-scale LP, or (2) select a few "good" scenarios so that one can cope with the arising LPs by the use of an ordinary 'commercial' direct method, or (3) something in between, e.g., embed an advanced sampling method into a well-tried decomposition method.

In order to treat more formally the alternate procedure consisting of observing random events and taking decisions, we state a mathematical formulation of a – slightly simplified – multistage stochastic program in the next section.

This facilitates the introduction of some technical terms and helps to make the subject accessible to readers who are not familiar with multistage stochastic programming. Afterwards, Section 3 states informally well-known existing techniques (with a focus on decomposition) and deals with an alternative approach designed to handle sequences of approximating programs. The paper ends up with some summarizing conclusions.

2. Multistage Stochastic Linear Program with Fixed Recourse

Using a dynamic programming type of recursion, we can write an MSLP in the following form:

for $t = H, \dots, 2$ define backwards

$$\begin{aligned} Q^t(x^{t-1}, \boldsymbol{\xi}^t(\omega)) := \min\ & \boldsymbol{c}^t(\omega)x^t(\omega) + E_{\boldsymbol{\xi}^{t+1}}[Q^{t+1}(x^t, \boldsymbol{\xi}^{t+1}(\omega))] \\ \text{s.t.}\ & W^t x^t(\omega) = \boldsymbol{h}^t(\omega) - T^{t-1}x^{t-1} \\ & x^t(\omega) \geqslant 0, \end{aligned} \tag{1}$$

with terminal value function given by

$$Q^{H+1}(\cdot, \cdot) \equiv 0.$$

The interesting minimal objective value and the today's optimal decision can then be obained by solving the problem

$$\begin{aligned} \min\ & z(x^1) := c^1 x^1 + \mathcal{Q}^2(x^1) \\ \text{s.t.}\ & W^1 x^1 = h^1 \\ & x^1 \geqslant 0. \end{aligned} \tag{2}$$

In case of unboundedness below or infeasibility of the program (1) $Q^t(x^{t-1}, \boldsymbol{\xi}^t(\omega))$ is set to $-\infty$ or $+\infty$, respectively. The *expectation functional* $\mathcal{Q}^2(\cdot)$ of the *value function* $Q^2(\cdot, \cdot)$ in (2) is defined as

$$\mathcal{Q}^2(x^1) := E_{P^{\boldsymbol{\xi}^2}}[Q^2(x^1, \cdot)] = \int_{\Xi^2} Q^2(x^1, \xi^2)\, \mathrm{d}P^{\boldsymbol{\xi}^2}(\xi^2) \tag{3}$$

$$= \int_{\Omega} Q^2(x^1, \boldsymbol{\xi}^2(\omega))\, \mathrm{d}P(\omega) = E_P[Q^2(x^1, \boldsymbol{\xi}^2)] \tag{4}$$

with

$$\begin{aligned} Q^2(x^1, \boldsymbol{\xi}^2(\omega)) := \min\ & \boldsymbol{c}^2(\omega)x^2(\omega) + \mathcal{Q}^3(x^2) \\ \text{s.t.}\ & W^2 x^2(\omega) = \boldsymbol{h}^2(\omega) - T^1 x^1 \\ & x^2(\omega) \geqslant 0. \end{aligned} \tag{5}$$

Sometimes the expectation functionals $\mathcal{Q}^t$ are named *expected value functions*, too (see, e.g., Kall and Wallace [6], p. 148). In the following, we refrain from a differentiation from 'normal' value functions; depending on the situation and notation it will become clear if it concerns the evaluation of an expectation or not. Occasionally, in literature the term *recourse function* is used instead of a value function; but the reader should be aware that sometimes *decision rules* or *policies* $\xi \mapsto x(\xi)$ are also named recourse functions (see, e.g., Wets [7], p. 566, or Rockafellar and Wets [8], S. 171).

In (3), Ξ^2 denotes the support of $\boldsymbol{\xi}^2$; concerning the definition of P, see below. Integrals in the form of $E_{P^{\xi^2}}[\cdot]$ are given by the sum of their positive and negative parts, with positive (negative) parts to be defined as $+\infty$ $(-\infty)$ when the integral diverges or the integrand takes the value $+\infty$ $(-\infty)$ on a set of strictly positive measure.[3] c^1 and h^1 are known vectors out of $\mathbb{R}^{n_1}$ and $\mathbb{R}^{m_1}$, respectively. For all stages $t = 2, \ldots, H$ the random vectors $\boldsymbol{\xi}^t := (\boldsymbol{c}^t, \boldsymbol{h}^t)$ fulfill

$$\begin{aligned} \boldsymbol{\xi}^t = (\boldsymbol{c}^t, \boldsymbol{h}^t)\colon\ (\Omega, \Sigma^t) &\longrightarrow (\mathbb{R}^{n_t+m_t}, \mathcal{B}^{n_t+m_t}) \\ \omega \quad &\longmapsto \boldsymbol{\xi}^t(\omega) = \begin{pmatrix} \boldsymbol{c}^t(\omega) \\ \boldsymbol{h}^t(\omega) \end{pmatrix} =: \begin{pmatrix} c^t \\ h^t \end{pmatrix} \end{aligned} \tag{6}$$

with $\Sigma^t \subseteq \Sigma^{t+1}$ σ-algebras relative to Ω $(t = 2, \ldots, H-1)$; herein, (6) stands for the $(\Sigma^t, \mathcal{B}^{n_t+m_t})$-measurability of $\boldsymbol{\xi}^t$.[4] Each σ-algebra Σ^t consists of those events that are 'known' at time t. Let P be a known probability measure on the measurable space (Ω, Σ^H). Because of the monotone increasing σ-algebras Σ^t – in the sense of set inclusion – the (Ω, Σ^t, P) are well-defined probability spaces for all $t = 2, \ldots, H$.

The distribution of $\boldsymbol{\xi}^t$ is assumed to be independent of past realizations. If this assumption is given up,[5] then the problem formulation must contain further information in the form of observed realizations of the random vectors up to time t. In order to keep the problems on a tractable level and to avoid an overloaded notation, herein we will assume stochastic independence between stages.

As to the decisions, we additionally demand the $(\Sigma^t, \mathcal{B}^{n_t})$-measurability of x^t for all t; this ensures the nonanticipativity of the decisions to be taken. x^{t-1} denotes the states of the system and $E_{P^{\xi^t}}[\cdot]$ the expected value w.r.t. the induced probability measure P^{ξ^t} $(t = 2, \ldots, H)$. The value functions $Q^t(\cdot, \cdot)$ and expectation functionals $\mathcal{Q}^t(\cdot)$ of stages $t > 2$ are defined analogously to $Q^2(\cdot, \cdot)$ and $\mathcal{Q}^2(\cdot)$, respectively; Ξ^t designates the support of $\boldsymbol{\xi}^t$.

Since the most common way to treat such a multistage stochastic linear program is to solve (approximating) discretized versions of the MSLP under consideration, the next section starts with the formulation of a deterministic equivalent program given a finite number of random outcomes. Afterwards, various types of solution techniques are stated rather verbally with a focus on well-known decomposition methods and on an alternative approach developed recently.

3. Solution Techniques

When deriving approximate problems by discretization, the multistage problem structure obviously results in the following effect: an increase of the number of decision stages leads to an exponential increase in problem size even if the number of realizations in each stage remains constant. To underline this effect, one could think of the following example: in a stochastic programming problem with 10 stages, 10 random variables per stage and only 10 random outcomes per stochastic dimension (i.e. a rather small sample of only 100 realizations of the 10-dimensional distribution on each stage, which can hardly obtain a sufficient estimation quality), one would have to deal with 10^{20} scenarios – a problem size which obviously is not tractable on a today's computer platform within acceptable time although the high-dimensional integration is reduced to a finite summation, as is shown in the following formulation of problem (1)–(2) in a discretized form:

$$
\begin{array}{lllr}
\min & c'x_0 + \sum_{s=1}^{S_1} p_1^s (c_1^s)'x_1^s + \sum_{s=1}^{S_2} p_2^s (c_2^s)'x_2^s + \cdots + \sum_{s=1}^{S_H} p_H^s (c_H^s)'x_H^s & & \\
\text{s.t.} & A\,x_0 & = b & \\
& T_1^s\,x_0 + W_1\,x_1^s & = h_1^s & \\
& & (s = 1, \ldots, S_1) & \\
& \qquad T_2^s\,x_1^{v(2,s)} + W_2\,x_2^s & = h_2^s & (7)\\
& & (s = 1, \ldots, S_2) & \\
& \qquad\qquad \ddots \qquad \ddots & \vdots & \\
& \qquad\qquad\qquad T_H^s\,x_{H-1}^{v(H,s)} + W_H\,x_H^s & = h_H^s & \\
& & (s = 1, \ldots, S_H) & \\
& x_0 \geqslant 0,\ x_t^s \geqslant 0, 1 \leqslant t \leqslant H,\ 1 \leqslant s \leqslant S_t. & &
\end{array}
$$

Herein, $x_0 := x_0^1$ and x_t^s $(t = 1, \ldots, H, s = 1, \ldots, S_t)$ indicate the today's decision and the decision in the future node (t, s), respectively.[6] $v(t, s)$ designates the state index of the antecedent node according to the underlying scenario tree, i.e. it holds $v(t, s) \in \{1, \ldots, S_{t-1}\}$ for all t, s.[7] But even for lower-dimensional problems with less stages, random variables and realizations, the above-mentioned *curse of dimensionality* has to be overcome by powerful solution techniques. As already said, up to a certain size of the DEP (7), commercial or academic direct methods can be applied successfully without taking advantage of the special problem structure, i.e. just using well-tried *sparse simplex methods* or interior point methods.

An alternative way to cope with larger DEPs is to exploit the special structure of multistage stochastic programs in the implementation of interior point algorithms.[8] For instance, using such an approach Yang and Zenios [39] were able to solve test problems with up to 2.6 million constraints and more than 18 million variables. In general, interior point algorithms turned out to be very effective to solve large-scale stochastic programming problems, likewise whether implemented to directly exploit problem structures or when applied to solve *master-* and *subproblems* within a decomposition method.

Because of the road of success of decomposition methods and their widespread usage, Subsection 3.1. briefly deals with the idea and realization of these approaches. Afterwards, Subsection 3.2. presents another kind of 'decomposing' a large-scale program into smaller subproblems within the handling of subsequent approximation steps.

3.1. Decomposition Methods

Considering a two-stage linear recourse problem with a finite number of second-stage realizations, then one can always form the two-stage extensive form. Its primal (dual) formulation has a block (block angular) structure, so that it seems natural to take advantage of this structure by performing a Benders [14] decomposition of the primal or a Dantzig–Wolfe [15] decomposition of the dual, respectively. Benders' method has been extendend in stochastic programming to take care of feasibility aspects and is known as *L-shaped method* by Van Slyke and Wets [16]: actually, it is a cutting plane technique which builds an outer linearization of the recourse function; the idea consists of sequentially adding *feasibility cuts* which determine the effective domain $\{x^1 \mid \mathcal{Q}^2(x^1) < +\infty\}$, and *optimality cuts* which are linear approximations to $\mathcal{Q}^2$ on its effective domain. For a detailed scheme of the L-shaped algorithm and enhancements like *multicut versions* or *bunching* of realizations, the interested reader is referred to Birge and Louveaux [5], Chapter 5. Furthermore, a variety of variants and extensions exist: e.g., adding a nonlinear regularized term to the objective (*regularized decomposition* by Ruszczyński [17]), or using sequential bounding approximations within the L-shaped method (see Birge [18] or Birge and Louveaux [5], pp. 296 ff).

As was already mentioned before, a potential disadvantage of sampling approaches is that some computational effort might be wasted on optimizing when the approximation is not accurate; an approach to avoid such difficulties is to combine sampling with another algorithm without complete optimization. An obvious candidate is to embed sampling into the L-shaped method. Two well-known examples of such a combined approach are given by (1) an importance sampling based approach by Dantzig and Glynn [19] and Dantzig and Infanger [20], and (2) the *stochastic decomposition* approach.[9] The first approach uses importance sampling to reduce variance when deriving each cut based on a large sample. In the second one, a single sample is employed to derive many cuts; the latter are less accurate and eventually drop away with increasing iteration number.

Generalizations of the classic two-stage L-shaped method to the multistage case have been done by Louveaux [22] who performed this extension for multi-stage quadratic problems, and by Birge [23] for the multistage linear case. The latter is also known as *nested L-shaped* or *nested Benders decomposition.* The basic idea is – analogously to the two-stage case – to place feasibility and optimality cuts; but now, optimality cuts are placed on the period t value function $\mathcal{Q}^{t+1}(x^t)$ for all t, and by feasibility cuts one achieves solutions x^t that have a feasible completion in all descendent scenarios. Again, the cuts represent consecutive (outer) linearizations of the value functions $\mathcal{Q}^{t+1}$; convergence to an optimal solution in a finite number of steps is due to the polyhedral convex structure of the $\mathcal{Q}^{t+1}$. However, despite of many similar features between the nested and the two-stage L-shaped method, there are some differences and peculiarities; we will only mention two of them:

(i) In the nested decomposition procedure, many alternative strategies are possible with regard to determining the next subproblem to solve, in particular on which stage t a subproblem will be solved. A so-called *fast-forward-fast-back*[10] sequencing protocol proposed by Wittrock [24] seems to be superior to other sequencing protocols like "fast-back" or "fast-forward" (see Gassmann [25] and Morton [26]).

(ii) The two-stage method always produces cuts which are supports of the value function $\mathcal{Q}^2$ if the subproblem is solved to optimality. In contrast to this, the nested method working with the above-mentioned "fast-forward-fast-back" protocol does not automatically generate a true support, i.e. cuts may lie strictly below the value function to be approximated.

Similar to the two-stage case, significant enhancement is possible for the nested method by efficient bunching (or *sifting*); see Birge [23] or Gassmann [25].

Decomposition methods which are based on an *Augmented Lagrangian* can be found in Ruszczyński [27], Rosa and Ruszczyński [28] or Mulvey and Ruszczyński [29]. For parallelization aspects concerning decomposition see, e.g., Ruszczyński [30].

It should be noticed, that a multistage stochastic program does not automatically demand for the application of the nested method. Even in the multistage case one can utilize the classic two-stage Benders decomposition as is done in Gondzio and Kouwenberg [31]: they divide 6-period problems into a "first stage" problem from today up to time $H_{\text{sub}} - 1$ and a set of "second stage" subproblems from time H_{sub} up to the horizon H. Applying the two-stage Benders decomposition which employs the interior point solver HOPDM[11] to optimize the master problem and the second stage subproblems, they were able to solve a problem with 25 million columns and 12.5 million rows in less than 5 hours – admittedly by massive parallelization using up to 13 processors.

On the one hand, this impressive result points out that the hope to solve truly large-scale deterministic equivalent linear programs became a certainty. On the other hand, this does not at all close the gap between demand and reality: the question still remains in how far this really helps to "solve" an original MSLP with a continuous underlying distribution. For instance, in the above-mentioned program with 25 million columns the corresponding scenario tree consisted of $13^6 = 4'826'809$ scenarios, namely a tree with 7 decision stages and 13 branches per node. Taking into account that 6 random coefficients are included in this special ALM model, the amount of 13 realizations does not appear that large. Of course, sophisticated methods can be employed to construct specific event trees which reproduce some properties of the underlying distributions; e.g., *tree-fitting* can be utilized to fit the first few moments of the underlying distributions (e.g., mean, standard deviation, skewness and kurtosis – depending on the number of realizations). But this only increases the confidence in the quality of the approximating discrete probability measure in the sense that one is "near" to the original distribution; one can at most expect that this helps to improve the quality of the solutions. It does not really help to quantify a solution quality in terms of probabilistic bounds (confidence intervals) or deterministic bounds on the true overall objective value. Statements on probabilistic bounds can only be given by statistical reasoning based on the sample size, and an evaluation of deterministic bounds requires special discrete support points. The next subsection deals with a specific discretization scheme – which has turned out to be an adequate technique given some structural properties of the multistage stochastic program are fulfilled – in conjunction with a numerical procedure to handle sequences of approximating deterministic equivalent linear programs: the choice of the discrete support points guarantees to evaluate upper and lower deterministic bounds on the true overall objective value, and the numerical procedure can cope effectively with the large-scale LPs which arise when one tries to successively improve these bounds.

3.2. Progressive Refinement Approach and Barycentric Approximation

Consider the general MSLP (1)–(2) stated in Section 2 with an underlying continuous joint distribution. When the problem has some – not very restrictive and rather technical – properties,[12] then the value functions $Q^t(x^{t-1}, \xi^t(\omega))$ in all stages are *saddle functions* with a convex behaviour w.r.t. x^{t-1} and h^t, and a concave behaviour w.r.t. c'. This saddle structure is exploited by the so-called *barycentric approximation scheme* in the following way: upper and lower approximations of the value functions at each stage are constructed leading to upper and lower scenario trees and associated large-scale LPs. The optimal objective values of these 'upper' and 'lower' LPs provide deterministic bounds on the true overall objective value of the original MSLP with its continuous distribution. Furthermore, they enable to state and localize the inaccuracy (in the sense of the distance between the upper and lower approximate objective values) w.r.t. the current approximations.

Usually, one's aim is to improve the quality of the approximations built up so far which is done by successive refinements of the support of the underlying distributions; in doing so, these supports are more and more discretized which leads to more ramified scenario trees and gives a better depiction of the future stochastic evolution. On balance, the procedure of successively raising the number of discretization points is inevitably linked to sequences of ever increasing[13] large-scale LPs which all have to be solved in order to track the development of the accuracy achieved.

Consider a scenario tree representing a deterministic equivalent program after having carried out a sequence of ν refinements of the support of the underlying probability distributions (where we confine ourselves to refinements in the root node only, i.e. partitioning always takes place w.r.t. the support of second-stage distributions). This refinement is restricted to the root node for two reasons: First, this kind of refinement results in a largest possible increase of the number of scenarios and thus represents the most challenging case since the corresponding LPs enlarge to the same extent. Secondly, this immense growth of the scenario tree stands for a very exhaustive incorporation of potential dynamics of the uncertain quantities and can be assumed to result in a strong improvement of the approximation accuracy.[14] The upper part of Figure 1 gives an exemplary image of such a refined scenario tree when the unrefined tree consisted of an ordinary binary tree. Having carried out a further refinement step $\nu + 1$ – in the sense that the chosen part of the support will be split – the subtree contained in the upper framed box will vanish and has to be replaced by new subtrees representing the new discretization points. If, for ease of simplicity, we think of a simple split of the partial support into two pieces, then the new parts are given by the two lower dashed boxes. Take note of the fact that, naturally, one can obtain new subtrees which are of exactly the same form as the vanishing part (for instance when each of the two new partial supports is represented by the same number of discretization points). We will come back to this fact when dealing with the need to solve these new subproblems on the operational level by means of a commercial LP solver. For every refinement step ν the scenario tree achieved is associated with a linear program LP_1^{ν} where the subscript 1 indicates the stage of the root node and the

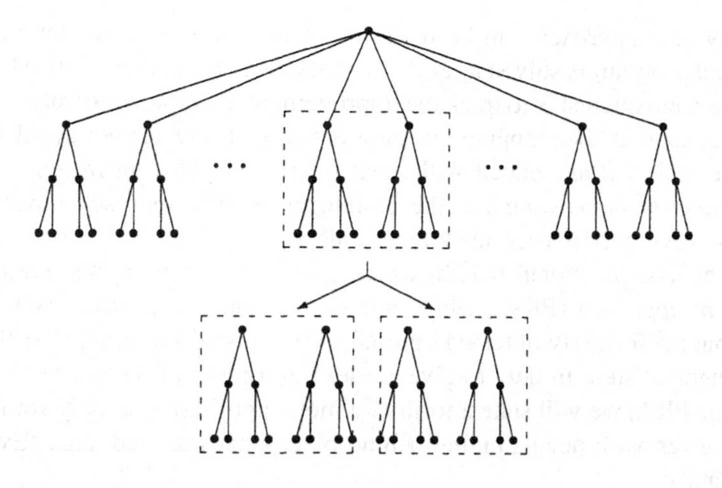

Figure 1. Changing parts of a scenario tree per refinement.

superscript ν the number of refinements carried out so far. Considering now the second-stage problems $LP_{2,s}^{\nu}$ (with $s = 1, \ldots, S^{\nu}$ both numbering the nodes on the second stage from left to right and reflecting the different states of the system on that stage, i.e. the specific discretization points of the support of ξ^2), the next refinement step $\nu + 1$ leads to the following situation in our specific example of Figure 1:

- The overall linear program LP_1^{ν} is replaced by $LP_1^{\nu+1}$. The new number of second-stage nodes is given by $S^{\nu+1} := S^{\nu} + 2$.
- The two old second-stage subproblems, say LP_{2,s^*}^{ν} and LP_{2,s^*+1}^{ν}, are replaced by $LP_{2,s^*}^{\nu+1}$,$LP_{2,s^*+1}^{\nu+1}$, $LP_{2,s^*+2}^{\nu+1}$ and $LP_{2,s^*+3}^{\nu+1}$.
- All the other old second-stage subproblems $LP_{2,s}^{\nu}, s \in \{1, \ldots, S^{\nu}\} \setminus \{s^*, s^* + 1\}$, remain unchanged and form the rest of the subproblems $LP_{2,s}^{\nu+1}$.

Usually, when employing whatever solver to optimize the large-scale LPs LP_1^{ν}, the following straightforward procedure is hardly suitable for coping with the refinement cycle:

1. Build up LP_1^0 and convey problem to the solver. Solve LP_1^0 and get optimal solution $x^{1,0}$. Set $\nu = 1$.
2. Carry out refinement step ν. Build up LP_1^{ν} and convey problem to the solver. Solve LP_1^{ν} and get optimal solution $x^{1,\nu}$. Set $\nu = \nu + 1$.
3. IF $\nu <$ Max[15] and $x^{1,\nu}$ does not fulfill a certain prechosen stopping-rule GOTO 2. ELSE STOP.

Naturally, this approach can be modified in the sense that the deterministic equivalent program is solved after every kth($k > 1$) refinement only. However, when the main interest is to track the improvement of single refinement steps or to stop as soon as the stopping criterion is fulfilled, one cannot avoid solving the large-scale LP associated with each refinement. Having in mind the so-called *curse of dimensionality*, the bottleneck of this approach obviously is given by solving these expanded large-scale LPs.

It is this computational burden which can be overcome by the *progressive refinement approach* (PRA), which reduces the numerical effort by trying to verify that the formerly obtained optimal solution is still optimal after the very last refinement step. In oder to give a brief description of the course of events within the PRA, we will state a form of a meta-algorithm (i.e. only some modules are given with details hiding behind black-boxes termed, e.g., 'Evaluate' or 'Update'):

Progressive Refinement Approach

1. Set $\nu = 0$.
2. Build up LP_1^ν and convey problem to the solver. Solve LP_1^ν and get optimal solution $x^{1,\nu}$.
   IF $\nu =$ Max or $x^{1,\nu}$ fulfills a certain prechosen stopping-rule STOP.
3. Solve all $LP_{2,s}^\nu$ with $x^{1,\nu}$ fixed in the RHS. Evaluate (sub-)gradient g^ν of z at $x^{1,\nu}$.
4. Set $\nu = \nu + 1$.
   Carry out refinement step ν. Build up new subproblems $LP_{2,s}^\nu$ and convey subproblems to the solver. Solve subproblems.
5. Update last evaluated (sub-)gradient $g^{\nu-1}$ to the current one g^ν. Check optimality, i.e. try to show that $-g^\nu \in \mathcal{N}_{X^1}(x^{1,\nu-1})$.[16]
   IF positive: Set $x^{1,\nu} = x^{1,\nu-1}$.
   IF $\nu <$ Max and $x^{1,\nu}$ does not fulfill a certain prechosen stopping-rule GOTO 4.
   ELSE STOP.
   ELSE (i.e. negative) GOTO 2.

It should be noticed that the optimality check in step 5 utilizes an only *sufficient* optimality condition, i.e. a negative optimality check does not mean that the optimal first-stage decision changes from refinement step $\nu \to \nu + 1$. Hence, a negative optimality check can cause a wasted effort to carry through steps 2 and 3, only to find out that the optimal first-stage decision did not change at all. Fortunately, when applying the PRA to two different multistage problem formulations (coming from the area of running financial products), the empirical results showed that this last-mentioned case occurs very seldom, if ever.

The refinement method utilized must necessarily possess some rather heuristic properties: with the ongoing refinement process, the replaced parts of the scenario tree have to become smaller and smaller relative to the whole scenario tree; additionally, all parts besides the replaced part have to remain totally unchanged (except the node probabilities of all second-stage nodes which are allowed to be changed). Otherwise the PRA cannot be expected to save much numerical effort and, hence, valuable computation time. The barycentric discretization scheme fits these requirements.

Altogether, the necessary handling of sequences of "refined" LPs – according to a barycentrically discretized support – could be shown to be manageable by the PRA with acceptable effort, at least with regard to two specific practical problems of running fixed-income products: depending on the problem characteristics (number of stages and random variables, lower or upper approximation, etc.) up to several hundreds of partitions of the second-stage support can be carried out within an elapsed CPU time of half an hour.[17] Quite contrary to the 'traditional' approach to solve every refined LP_1^ν as a whole: there, only a few refinement steps are possible within a comparable period of time.

The main advantage is given by a very small number of new subproblems $LP_{2,s}^\nu$ which additionally have the same problem dimensions, so that within the simplex algorithm an effective start from an advanced basis can be employed. Because of the (mostly) small number of subproblems to be solved in a refinement step – and because of the hot-start possibility keeping the simplex algorithm time for the subproblems in an order of magnitude of some seconds, parallelization is not so essential for a successful handling of the refinement cycle. Of course, to solve the subproblems in parallel could be implemented without further ado. However, we assume that a real acceleration of the computation will only be achieved in case of a negative optimality check (i.e. when the solution formerly achieved could not be verified to be still optimal after the very last refinement step) because then all subproblems arisen so far have to be solved again;[18] but the latter situation was only rarely observed in the two problem statements considered. Comprehensive numerical results with regard to CPU time and a more detailed presentation of the practical problems (funding conventional Swiss mortgages and investing savings account deposits) can be found in Haarbrücker [37], and Frauendorfer and Haarbrücker [38].

4. Summary

A variety of suitable methodologies, numerical techniques and optimization tools exists to deal with multistage stochastic programs. For instance, deterministic equivalent programs – corresponding to approximating or originally discrete probability measures – can be solved with well-tried direct methods (commercial or academic solvers) or sophisticated indirect methods (decomposition, stochastic decomposition, Lagrangian-based approaches). In order to achieve such approximating discrete probability measures and associated scenario trees, a multitude of discretization procedures can be applied: sampling (Monte Carlo – possibly in conjunction with variance-reducing techniques, EVPI-based importance sampling, and many others), selecting some kind of "good" scenarios" providing deterministic bounds (e.g., barycentric approximation, second-order scenario approximation), generating representative scenarios, and so on. In some cases, sampling/discretization procedures can be embedded in or at least combined with the above-mentioned direct or indirect methods which often leads to efficient numerical techniques. Another way to decrease computation time is given by techniques which aim at smaller scenario trees, like scenario aggregation or scenario reduction techniques.

As has been already mentioned, when one utilizes such a sophisticated method of a piece or combines some good modules in an efficient way, it is quite possible, for instance, to solve very large-scale deterministic equivalent programs with up to 25 million columns and nearly 13 million rows in a parallelized manner in a few hours, or to handle sequences of hundreds of successively discretized programs.

On balance, two aspects play – and will play! – a key role to cope successfully with MSLPs which model practical problem situations (i.e. rather high- than low-dimensional problems): to keep the gap between the original and the approximate distributions as close as possible, and in the meantime to keep an eye on the numerical solvability of the corresponding large-scale LPs.

Notes

1. Still to be defined!

2. With regard to the original continuous joint distribution.

3. Cases of simultaneous occurrence of $-\infty$ and $+\infty$ are handled by the convention $(-\infty)+\infty = +\infty$ (see, e.g., Wets [9], p. 312), i.e. infeasibility in one subproblem cannot be compensated by unboundedness below in another subproblem.

4. $\mathcal{B}^n (n \in \mathbb{N}$ designates the Borel σ-algebra on $\mathbb{R}^n$, i.e. the σ-algebra which is generated by all open subsets of $\mathbb{R}^n$.

5. Take note of the fact that relaxing the stochastic independence cannot be done without any compensation but only in favour of some weaker kind of regularity condition like, e.g., linear dependency (see Wets [10], pp. 202f, 205).

6. The stage index $t = 1, \ldots, H$ is now written as subscript in order to simplify the notation.

7. With s used as state index, S_t indicating the total number of nodes (states) on stage t, and p_t^s denoting the probability of node (t, s).

8. Since a detailed treatise of this subject is out of the scope of this paper, the reader is referred to, e.g., Birge and Qi [11], Yessup, Yang and Zenios [12], or Czyzyk, Fourer and Mehrotra [13].

9. See Higle and Sen [21].

10. I.e. the algorithm proceeds from stage t in a specific direction – either $t + 1$ ("forward") or $t - 1$ ("back") until it can no longer proceed in that direction.

11. See Gondzio [32].

12. For a detailed treatise see, e.g., Frauendorfer [33–35].

13. At least, if no *scenario aggregation* procedures are applied or "important" scenarios are selected.

14. E.g., as to the barycentric approximation scheme, this latter reasoning can be found in detail in Marohn [36], Chapter 5.

15. Max indicating the maximal number of refinement steps.

16. $\mathcal{N}_{X^1}(x^{1,\nu-1})$ denotes the normal cone to the feasibility region X^1 of x^1 at the previously achieved optimal solution $x^{1,\nu-1}$.

17. Even on a small Unix workstation with a moderate memory of 256 MB RAM.

18. Regardless of the difficulty that in those cases the (presumably very large) overall linear program LP_1^ν has to be solved, too.

Bibliography

[1] Birge, J. R. and Wallace, S. W.: A separable piecewise linear upper bound for stochastic linear programs, *SIAM J. Control Optim.* **26** (1988), 725–739.

[2] Birge, J. R., Wets and R. J.-B.: Sublinear upper bounds for stochastic programs with recourse, *Math. Programming* **43** (1989), 131–149.

[3] Wallace, S. W.: Solving stochastic programs with network recourse, *Networks* **16** (1986), 295–317.

[4] Birge, J. R. and Qi, L.: Continuous approximation schemes for stochastic progams, *Ann. Oper. Res.* **56** (1995), 15–38.

[5] Birge, J. R. and Louveaux, F.: *Introduction to Stochastic Programming*, Springer-Verlag, New York, 1997.

[6] Kall, P. and Wallace, S. W.: *Stochastic Programming*, Wiley, Chichester, 1994.

[7] Wets, R. J.-B.: Stochastic programming: Solution techniques and approximation schemes, In: A. Bachem, M. Grötschel and B. Korte (eds), *Mathematical Programming: The State-of-the-art 1982*, Springer-Verlag, Berlin, 1983, pp. 566–603.

[8] Rockafellar, R. T. and Wets, R. J.-B.: Nonanticipativity and $\mathcal{L}^1$-martingales in stochastic optimization problems, *Math. Programming Study* **6** (1976), 170–187.

[9] Wets, R. J.-B.: Stochastic programs with fixed recourse: The equivalent deterministic program, *SIAM Rev.* **16** (1974), 309–339.

[10] Wets, R. J.-B.: Stochastic programs with recourse: A basic theorem for multistage problems, *Z. Wahrschein. verw. Geb.* **21** (1972), 201–206.

[11] Birge, J. R. and Qi, L.: Computing block-angular Karmarkar projections with applications to stochastic programmin, *Management Sci.* **34** (1988), 1472–1479.

[12] Jessup, E. R., Yang, D. and Zenios, S. A.: Parallel factorization of structured matrices arising in stochastic programming, *SIAM J. Optim.* **4** (1994), 833–846.

[13] Czyzyk, J., Fourer, R. and Mehrotra, S.: Using a massively parallel processor to solve large sparse linear programs by an interior point method, *SIAM J. Sci. Comput.* **19** (1998), 553–565.

[14] Benders, J.: Partitioning methods for solving mixed variables programming problems, *Numer. Math.* **4** (1962), 238–252.

[15] Dantzig, G. B. and Wolfe, P.: The decomposition principle for linear programs, *Oper. Res.* **8** (1960), 101–111.

[16] Van Slyke, R. M. and Wets, R. J.-B.: L-shaped linear programs with application to optimal control and stochastic programming, *SIAM J. Appl. Math.* **17** (1969), 638–663.

[17] Ruszczyński, A.: A regularized decomposition method for minimizing a sum of polyhedral functions, *Math. Programming* **35** (1986), 309–333.

[18] Birge, J. R.: Using sequential approximations in the L-shaped and generalized programming algorithms for stochastic linear programs, Technical Report 83-12, Department of Industrial and Operations Engineering, University of Michigan, Ann Arbor, MI, 1983.

[19] Dantzig, G. B. and Glynn, P.: Parallel processors for planning under uncertainty, *Ann. Oper. Res.* **22** (1990), 1–21.

[20] Dantzig, G. B. and Infanger, G.: Large-scale stochastic linear programs – Importance sampling and Benders decomposition, In: *Computational and Applied Mathematics, I. Algorithms and Theory, Sel. Rev. Pap. IMACS 13th World Congr., Dublin/Irel. 1991*, 1992, pp. 111–120.

[21] Higle, J. and Sen, S.: Stochastic decomposition: An algorithm for two stage linear programs with recourse, *Math. Oper. Res.* **16** (1991), 650–669.

[22] Louveaux, F.: A solution method for multistage stochastic programs with recourse with application to an energy investment problem, *Oper. Res.* **28** (1980), 889–902.

[23] Birge, J. R.: Decomposition and partitioning methods for multistage stochastic linear programs, *Oper. Res.* **33** (1985), 989–1007.

[24] Wittrock, R. J.: Advances in a nested decomposition algorithm for solving staircase linear programs, Technical Report SOL 83-2, Systems Optimization Laboratory, Stanford University, Stanford, CA, 1983.

[25] Gassmann, H.: MSLIP, a computer code for the multistage stochastic linear programming problem, *Math. Programming* **47** (1990), 407–423.

[26] Morton, D. P.: An enhanced decomposition algorithm for multistage stochastic hydroelectric scheduling, Technical Report NPSOR-94-001, Department of Operations Research, Naval Postgraduate School, Monterey, CA, 1994.

[27] Ruszczyński, A.: An augmented Lagrangian decomposition method for block diagonal linear programming problems, *Oper. Res. Lett.* **8** (1989), 287–294.

[28] Rosa, C. and Ruszczyński, A.: On augmented Lagrangian decomposition methods for multistage stochastic programs, Working Paper WP-94-125, IIASA International Institute for Applied Systems Analysis, Laxenburg, Austria, 1994.

[29] Mulvey, J. M. and Ruszczyński, A.: A new scenario decomposition method for large-scale stochastic optimization, *Oper. Res.* **43**(3) (1995), 333–353.

[30] Ruszczyński, A.: Parallel decomposition of multistage stochastic programming problems, *Math. Programming* **58** (1993), 201–228.

[31] Gondzio, J. and Kouwenberg, R.: High performance computing for asset liability management, Preprint MS-99-004, Department of Mathematics & Statistics, The University of Edinburgh, UK, 1999.

[32] Gondzio, J.: HOPDM (version 2.12) – A fast LP solver based on a primal-dual interior point method, *Europ. J. Oper. Res.* **85** (1955), 221–225.

[33] Frauendorfer, K.: *Stochastic Two-Stage Programming*, Lecture Notes in Econom. Math. Systems 392, Springer-Verlag, Berlin, 1992.

[34] Frauendorfer, K.: Multistage stochastic programming: Error analysis for the convex case, *Z. Oper. Res.* **39**(1) (1994), 93–122.

[35] Frauendorfer, K.: Barycentric scenario trees in convex multistage stochastic programming, *Math. Programming* **75**(2) (1996), 277–294.

[36] Marohn, C.: *Stochastische mehrstufige lineare Programmierung im Asset & Liability Management*, Bank- und finanzwirtschaftliche Forschungen, Bd. 282, Paul Haupt Verlag, Bern, 1998.

[37] Haarbrücker, G.: *Sequentielle Optimierung verfeinerter Approximationen in der mehrstufigen stochastischen linearen Programmierung*, Doctoral thesis No. 2410, University of St. Gallen, 2000.

[38] Frauendorfer, K. and Haarbrücker, G.: Solving sequences of refined multistage stochastic linear programs, Submitted for appearance in *Ann. Oper. Res.*

[39] Yang, D. and Zenios, S. A.: A scalable parallel interior point algorithm for stochastic linear programming and robust optimization, *Comput. Optim. Appl.* **7**(1) (1997), 143–158.

Chapter 8

ON THE EFFICIENCY AND EFFECTIVENESS OF CONTROLLED RANDOM SEARCH*

Eligius M. T. Hendrix
Group Operations Research and Logistics
Wageningen University, Hollandseweg 1
6706 KN Wageningen, The Netherlands

Pilar M. Ortigosa and Inmaculada García
Computer Architecture & Electronics Dpt.
University of Almería, Cta. Sacramento SN
04120 Almería, Spain

Abstract Applying evolutionary algorithms based on populations of trial points is attractive in many fields nowadays. Apart from the evolutionary analogy, profound analysis on their performance is lacking. In this paper, within a framework to study the behaviour of algorithms, an analysis is given on the performance of Controlled Random Search (CRS), a simple population based Global Optimization algorithm. The question is for which functions (cases) and which parameter settings the algorithm is effective and how the efficiency can be influenced. For this, several performance indicators are described. Analytical and experimental results on effectiveness and speed of convergence (Success Rate) of CRS are presented.

Keywords: Controlled Random Search, speed of convergence, smooth optimization, stochastic algorithms, evolutionary algorithms, effectiveness

1. Introduction

Evolutionary algorithms are applied nowadays in many fields such as hydrology, electrical engineering, food science etc. In optimization problems where many local minima may occur, the idea is that the stochastic population based algorithms called genetic algorithms and evolutionary algorithms

*This work was supported by the Ministry of Education of Spain (CICYT TIC99-0361).

G. Dzemyda et al. (eds.), Stochastic and Global Optimization, 129–145.
© 2002 *Kluwer Academic Publishers.*

scan the parameter area efficiently legitimated with the rationale of the analogy of evolution. Many publications in those applied fields report on running a few variants of the algorithms on one or several practical problems. Would other algorithms from classical nonlinear optimization or other stochastic or deterministic Global optimization ideas have performed as good or better for the practical application?

Research on the performance of optimization methods focuses on the question when to apply which type of algorithms. If we want to investigate the behaviour of algorithms systematically, one should measure performance depending on parameters of the algorithm and characteristics of the optimization problem in a structured way. Therefore, in [1] a framework was suggested for the investigation of algorithms. For the algorithm we are investigating the framework is depicted in Figure 1.

It is a challenge to investigate properties of population based algorithms further than the evolutionary analogy and the usual so-called infinity-effort property of stochastic algorithms; i.e. that if the algorithms are proceeded until infinity, the global optimum will be found with certainty. The target of this paper is to investigate within the framework, the efficiency and effectiveness of a simple population algorithm called Controlled Random Search. For this it is necessary to describe the parameters of the algorithm, the characteristics of designed or used test problems and to define more precisely what is effectiveness and efficiency, preferably with some measurable performance indicators. Therefore, in this section a description of the algorithm is given and various ideas about the performance measurement are discussed. In the following sections, analytical and experimental results are provided.

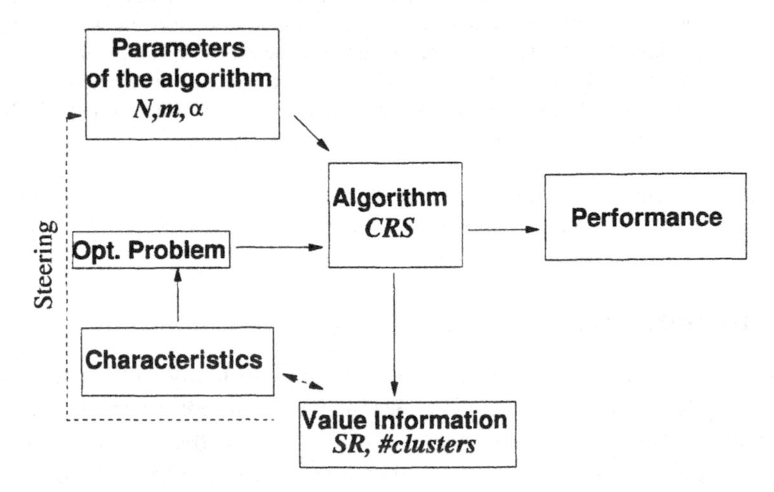

Figure 1. View on investigating stochastic Global Optimization Algorithms for the CRS algorithm.

1.1. The Algorithm under Study, CRS

The Controlled Random Search algorithm, proposed by Price [2–4], is a simple and direct procedure for global optimization. It has shown to be applicable both to unconstrained and constrained optimization problems [5,6]. In the analysis the global optimization problem to be solved is that described by (1), where X is a hyper-rectangle.

$$\min f(x), x \in X \subset R^n. \tag{1}$$

CRS starts by filling a population R initially with a sample of N trial points uniformly distributed over the search space X. The algorithm iteratively updates the set of trial points (population) R until some stopping conditions are reached. At every iterative step two different kinds of trial points may be computed; the so called primary and secondary trial points. Both kinds of trial points are generated by performing an operation on a subset $\{r_1, \ldots, r_{m+1}\} \subset R$ of $m + 1$ trial points. The points r_i, $i = 1, \ldots, m + 1$, of this generation subset are randomly selected from the current population R of N points. In the first versions of the algorithm [2–4], m was taken as the dimension n. Primary points are generated in a Nelder–Mead fashion [7] by mirroring (reflecting) a point (r_{m+1}) over the centroid, $\overline{G}$, of the remaining subset of points $r_1, \ldots, r_m$. A secondary point is located in the middle between r_{m+1} and the centroid $\overline{G}$ (see Figure 2). While primary points are intended to keep the search space wide (global search), secondary points are conductive to convergence (local search).

Secondary points are only computed if the current primary trial point fails and the estimate (sr_k) of the success rate on finding better function values than the worst value y_k in the current population is below the value of a parameter α (Price suggested $\alpha = 0.5$). This general procedure may be modified in a variety of ways. The investigated version of CRS is detailed at Algorithm 1.

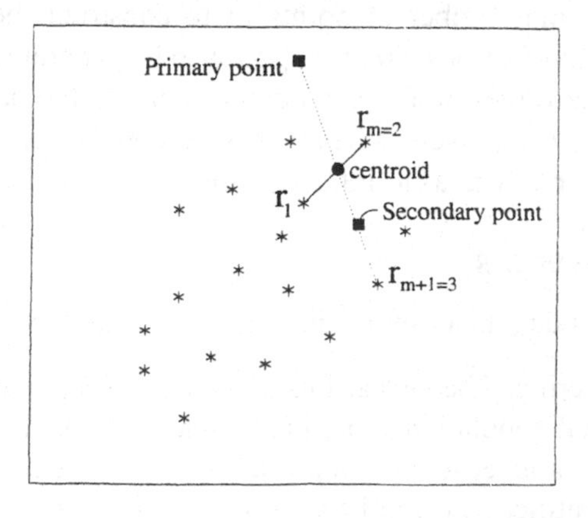

Figure 2. Controlled Random Search.

ALGORITHM 1. **CRS**(f, X, N, m, α)

Set $k = 0$ and $ns = 0$
Generate and evaluate a set R, of N random points uniformly on X
while (Stopping Conditions) # **e.g.** $\max f(r_i) - \min f(r_i) < \varepsilon$.
  select at random a subset $\{r_1, \dots, r_{m+1}\}$ from R
  $\overline{G} = \frac{1}{m}\sum_{i=1}^{m} r_i$
  $\overline{P} = 2 \times \overline{G} - r_{m+1}$ # **Primary points**
  $k = k + 1$
  determine $y_k = f(r^{max}) = max_i f(r_i), r_i \in R$
  if $\overline{P} \in X$ AND $f(\overline{P}) < y_k$
    replace r^{max} with $\overline{P}$ in R; $ns = ns + 1$
  else if $(sr_k = ns/k) < \alpha$ # **Rate of Success Test**
    $\overline{P} = \frac{1}{2}(\overline{G} + r_{m+1})$ # **Secondary points**
    if $f(\overline{P}) < y_k$
      replace r^{max} with $\overline{P}$ in R; $ns = ns + 1$
    $k = k + 1$
End while

Note that whenever the population has reached a convex level set (or one convex compartment of a level set), the calculation of a secondary point implies an improvement (function value is not worse than y_k). However, this does not imply that the algorithm starts behaving in a constant improving manner due to the secondary points. Namely, when the Success Rate exceeds the value of α, only Primary points are generated. This means that in the analysis of the Success Rate we should take care when it starts approaching the level of α. Around this level, the algorithm fluctuates in its decision of generating secondary points.

Parameters of the algorithm which can be influenced by the user are the population size N, the number of points m to construct the centroid and α which controls the number of secondary points to be generated. The interesting questions is how the values of the parameters influence the performance of the algorithm and how a user, given value information on the problem to be solved, can set the parameter values as to influence the search for the optimum.

1.2. Effectiveness

There are several targets a user of the algorithm may have:

1. Uniform covering: The idea as introduced by Klepper and Hendrix [8,9], is that the final population should resemble a sample from a uniform distribution over a level set with a predefined level. The practical relevance is due to identification problems in parameter estimation. Performance indicators are difficult to construct. Usually one partitions the final re-

gion in subsets and sums the deviations from the expected number of points in each partition set.

2. To discover all global minimum points. This of course can only be realised when the number of global minimum points is finite. Performance indicators are based on the convergence towards all global minimum points.

3. To detect at least one global optimal point. A performance indicator can be defined as: How many times over several random series and/or several test functions has the minimum point been detected. Standard test functions (see [10]) or constructed cases can be used. In Section 4, the Hitting Rate (HR) performance indicator will be defined and computed to experimentally evaluate the CRS algorithm.

In this paper we focus on the second and third target.

1.3. Efficiency

Globally efficiency is defined as the effort the algorithm needs to reach the target. A usual indicator in stochastic algorithms is the number of function evaluations necessary to reach the optimum. However, this indicator depends on many factors such as the shape of the test function and the termination criteria used. Therefore we focus on an indicator called Success Rate, SR, as described by Hendrix *et al.* [1], which is close to the concept of convergence speed; it only measures the limit behaviour of the algorithm. The relevance for the concept of convergence speed is due to the analyses by Zabinsky and Smith [11] and Baritompa *et al.* [12], who show that a fixed success rate of an effective algorithm in the sense of uniform covering would lead to an algorithm which is polynomial in the dimension of the problem, i.e. the expected number of function evaluations grows polynomially with the dimension of the problem. Therefore we are particularly interested in the behaviour of the Success Rate when the dimension of the problem increases. For the analysis two major cases can be distinguished:

1. Success rate when one global minimum point is reached.

2. Success rate when several global minimum points are reached.

In Section 2, case 1 is investigated for smooth non-linear optimization problems. In Sections 3 and 4 the aspect of effectiveness is addressed.

2. Efficiency when One Global Minimum Point Is Reached

In smooth non-linear optimization the convergence speed is usually related to the limit behaviour of algorithms on a final ellipsoidal level set. If we focus on the success rate for smooth functions, i.e. with an ellipsoidal limit level set, the question is how the success rate depends on the shape of the level set around the minimum point. In [1] an analysis can be found on the SR for increasing values of the problem dimension from the point of view of complexity and an analytical prediction is confronted with the empirical measurement for spherical level sets.

Here we present the results of a small experiment. For the case when one global minimum point is reached in smooth optimization, the optimization problem is simply characterised by the condition number of the Hessean in the minimum point. Therefore in a two-dimensional case it is sufficient to describe the test problem as:

$$f(x) = x_1^2 + L \cdot x_2^2. \tag{2}$$

Values of SR were experimentally obtained for the elliptical problem as defined by Equation (2) with $L = 1.0$ (two-dimensional spherical problem), $L = 10^{-1}$ and $L = 10^{-3}$ (two-dimensional elliptical functions). As it follows from [1] that the result (SR) does not depend on N, we chose to vary the other parameters α and m here. In total, we create an experimental design with $\alpha = 0.0, 0.1, \ldots, 0.9, 1.0$ and with $m = 2$ and $m = 10$. Figure 3 shows the results.

Let us first focus on the "flatness" of the final ellipse. It can be derived that when the problem is scaled with the parameter L, one obtains a spherical problem ($L = 1.0$), for which the same random series would result in exactly the same development of the algorithm as for the non-scaled problem (2). This can be observed from Figure 3 which clearly shows that the speed of convergence of CRS does not depend on L, which describes the shape of the final level set. It is known from the analysis in [1] that a higher value of m, leads to concentration of the centroid around the middle of the level set, giving a better success rate than for low values of m. This effect can also be observed from Figure 3.

With respect to parameter α, one should realise that secondary points always result in an improvement, i.e. they generate convex combinations of the parents which leads to improvements for convex (elliptic) limit level sets. Here we try to quantify the improvement of SR by changing the parameter α. For values of α, lower than the SR of primary points, secondary points are never generated, so the success rate is not influenced by α. For high values of α, a secondary point is generated at every iteration where the primary point fails and the resulting success rate is a fixed average of both success rates. In between, one can observe that on a part of the curve the success rate equals the value for α. In those cases the sr_k indicator of the algorithm remains fluctuating around the value of α. When sr_k is below α, the secondary points help the sr_k to get a value which exceeds α. When it has reached this level only primary points are

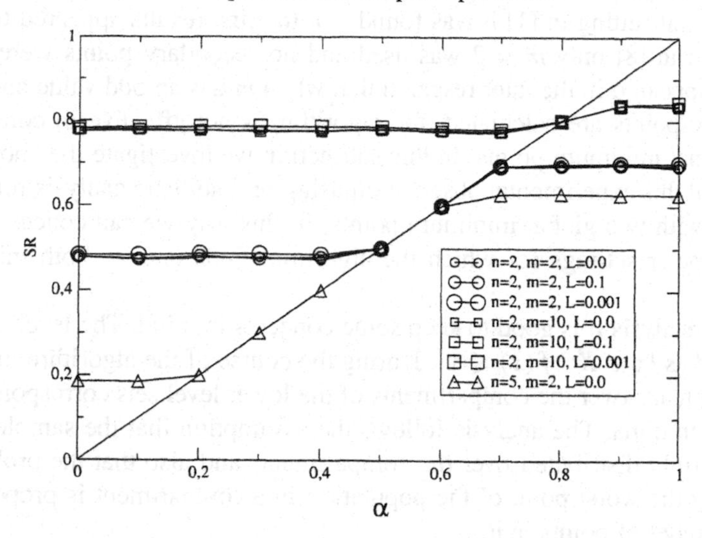

Figure 3. Values of SR versus α, experimentally obtained for a Spherical and two different Elliptical functions ($L = 0.1$, $L = 0.001$).

calculated which lower the value of sr_k in the direction of the success rate of primary points.

The exercise in this section shows how the framework of Figure 1 can be used to create an experimental design where performance indicators are defined and measured and parameters and characteristics of cases are varied in a systematic way to verify or invalidate hypotheses on the behaviour of stochastic algorithms.

3. Effectiveness in Reaching All Global Minimum Points

For the question of effectiveness, we investigate in Subsection 3.1. in an analytical way, based on the simple example of a bispherical function, the second target: in which cases does the algorithm converge to several global minimum points? We can use the insight in the mechanism to understand the behaviour of the algorithm with respect to the first target: in which cases does CRS detect one global minimum point?

Later, in Subsection 3.2. some experiments are done based on variants with two global optima of which compartments of the level set differ in size or shape.

3.1. Analytical Observations

In the results of [8] where the focus was on uniform covering, it was concluded from some simple test cases that CRS does surprisingly well in covering the global minimum points. Much later however, in a more extensive experimental setting in [1] it was found that the first results appeared to be an exception; in [8] only $m = 2$ was used and no secondary points were calculated. It appeared in the later research that when m has an odd value and when secondary points are calculated, the algorithm is not effective in converging to all global minimum points. In this subsection we investigate the "how" and "when" of this experimental observation using probabilistic analysis for a simple case with two global minimum points. In this way we can concentrate on the parameter settings for which the algorithm finds one, or both minimum points.

In the analysis it is good to keep some concepts in mind. The level set $S(y)$ is defined as $\{x \in X\colon\ f(x) \leqslant y\}$. During the course of the algorithm, the population spreads over the compartments of the lower level sets corresponding to the local minima. The analysis follows the assumption that the sample points are uniformly distributed over the compartments and also that the probability of finding the worst point of the population in a compartment is proportional to the number of points in it.

In the first studied case which is called the *bispherical problem* the compartments around the two global minimum points have the same size:

$$f(x) = \min \begin{cases} (x_1 - 0.5)^2 + x_2^2, \\ (x_1 + 0.5)^2 + x_2^2. \end{cases} \tag{3}$$

For the analysis a state variable P is introduced which is the fraction of points which can be found in the first compartment; $P = N_1/N$ where N_1 is the number of points (also called a cluster) in the first compartment. Assume (w.l.o.g.) $P > 1/2$. The questions now are: in which cases does $P \to 1$ and when does $P \to 1/2$. The analysis focuses on the probability that P (and so N_1) increases and the probability it decreases. To understand the mechanism, it is useful to put Figure 4 in mind. Because secondary and primary points produce different convergence effects they will be analysed separately.

Secondary Points. They only give a change in the state of the population when there is an improvement, i.e. situation 1 of Figure 4; $m + 1$ points are in the same compartment (cluster). An increase of N_1 takes place when $r^{\max}$ (the worst point in the cluster) is in cluster 2 (probability $= (1 - P)$) and the $m + 1$ points are in cluster 1:

$$P(\text{increase } N_1|\text{sec. point}) = P^{m+1} \times (1 - P),$$
$$P(\text{decrease } N_1|\text{sec. point}) = (1 - P)^{m+1} \times P.$$

In all other cases no change of N_1 occurs. Investigating the ratio of those probabilities shows us that:

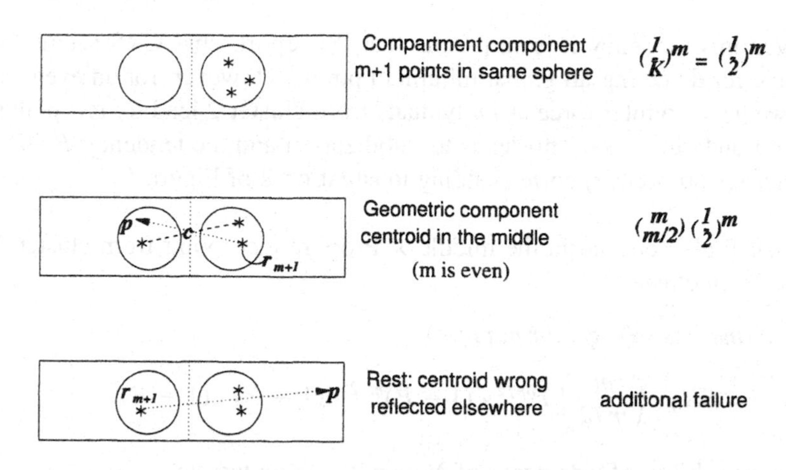

Figure 4. Various possibilities in the course of the algorithm.

$$\frac{P^{m+1} \times (1-P)}{(1-P)^{m+1} \times P} = \left(\frac{P}{1-P}\right)^m > 1 \quad \text{for } P > 1/2.$$

From this we can conclude that secondary points lead to convergence to the compartment where most of the points are. More generally, when there are several global minimum points, initially the compartment with the biggest size will contain the largest amount of points (biggest P). Secondary points have the tendency to move points from smaller compartments to the largest one. In smooth optimization the compartments of the minimum points are in limit defined by ellipsoids determined by the eigenvalues of the Hessean. The minimum point with the compartment with the largest size (volume) is defined by the lowest determinant of the Hessean.

Primary points. The analysis of the tendency for primary points is slightly more complicated. We are not only dealing with situation 1 of Figure 4 which was called the *compartment component* in [1], but for even value of m the centroid may end up in the middle leading to situation 2 in Figure 4, called the *geometric component.* In contrast to secondary points, primary points in situation 1 (and 2) do not always lead to improvements. The probability SR_0 of an improvement is the Success Rate of the spherical problem. When studying the ratio between probability of increase of P and probability of decrease of P, this Success Rate of course does not matter.

The compartment component gives a similar tendency as the secondary points, namely $P \rightarrow 1$:

$$P(increaseN_1, comp.prim.point) = SR_0 \times P^{m+1} \times (1-P),$$
$$P(decreaseN_1, comp.prim.point) = SR_0 \times (1-P)^{m+1} \times P.$$

Given this tendency it is surprising that [8] reports that CRS seems to be effective for detecting all global minimum points. However, for an even value of m we get a counter force as individuals from cluster 2 lead to new points in cluster 1 and viceversa. This leads to stabilization and the tendency $P \to 1/2$ (geometric component) corresponding to situation 2 of Figure 4:

$SR_0\times$ Prob. centroid in the middle $\times$ Prob. reflect. point from cluster-2 $\times$ Prob. $r^{\max}$ in cluster-2:

$$P(increase\, N_1, geo.prim.point) = SR_0 \cdot \binom{m}{m/2} P^{m/2} \cdot (1-P)^{m/2} \cdot (1-P) \cdot (1-P).$$

The probability of a decrease of N_1 can be computed as:

$SR_0\times$ Prob. centroid in the middle $\times$ Prob. reflect. point from cluster-1 $\times$ Prob. $r^{\max}$ in cluster-1:

$$P(decrease\, N_1, geo.prim.point) = SR_0 \cdot \binom{m}{m/2} P^{m/2} \cdot (1-P)^{m/2} \cdot P \cdot P.$$

So for m even and $P > 1/2$ the decrease is bigger than increase for the geometric part leading to a tendency $P \to 1/2$. What about the total? For m even, the ratio between probability of increase and decrease, given we are generating a primary point is:

$$\frac{P^{m+1} \cdot (1-P) + \binom{m}{m/2} \cdot P^{m/2} \cdot (1-P)^{m/2} \cdot (1-P)^2}{P \cdot (1-P)^{m+1} + \binom{m}{m/2} \cdot P^{m/2} \cdot (1-P)^{m/2} \cdot P^2}. \tag{4}$$

In particular for $m = 2$ (used in [8]) Equation (4) becomes:

$$\frac{P^2 + 2 \cdot (1-P)^2}{(1-P)^2 + 2 \cdot P^2} < 1 \quad \text{for } P > 1/2. \tag{5}$$

Independent of the initial value of P, we have convergence to $P = 1/2$; i.e. if $P > 1/2$, it decreases, when $P < 1/2$ it increases. However, evaluating ratio (4) for bigger values of m (even) leads to the conclusion that the convergence depends on the starting value of P. For instance for an initial value of $P = 0.6$, ratio (4) is smaller than 1 up to $m = 14$ and the algorithm is effective in finding both minimum points. However for $P = 0.9$ most of the points are clustered around the first minimum point and ratio (4) is bigger than 1 such that only one minimum point is reached.

Concluding we have found an analytical basis of the effectiveness reported by Klepper and Hendrix [8] for the bispherical problem running CRS with

$m = 2$ without secondary points. When we deviate from the parameter values and characteristic of this case, the following happens:

1. For odd values of m there is no geometrical component and the algorithm converges to one minimum point.
2. Using secondary points gives a stronger tendency (and speed) of reaching one minimum point.
3. For even and higher (than $m = 2$) values for m, the effectiveness depends on the initial value of P.
4. When both compartments are not as big or when both compartments are elliptic, i.e. the eigenvalues of the Hessean differ, the geometric component is less and the algorithm looses its effectiveness. This is experimentally investigated in Subsection 3.2..

3.2. Experimental Effectiveness Result for Bispherical and Bielliptical Problems

CRS(N, m, α) was experimentally investigated for two functions: a bispherical function and a bielliptical function defined by Equation (6). Both functions have two global minimum points and the parameter γ determines the relative size of the two final level sets. For the bielliptical function the relative orientation and center location are given by ϕ, and a and b, respectively. The bispherical function corresponds to parameter values $\phi = 0.0$, $a = 1.0$, $b = 0.0$, $L_0 = L_\phi = 1.0$ and the bielliptical to $\phi = \pi/6$, $a = 0.3$, $b = -0.3$, $L_0 = L_\phi = 10.0$.

$$f(x) = \min \begin{cases} x_1^2 + L_0 \cdot x_2^2, \\ \gamma \cdot [((x_1 - a)\cos\phi + (x_2 - b)\sin\phi)^2 + \\ \quad + L_\phi((x_1 - a)\sin\phi - (x_2 - b)\cos\phi)^2]. \end{cases} \tag{6}$$

Tables 1 and 2 provide experimental results on the number of times both global minimum points are detected for the bispherical and bielliptical functions. For the bispherical function we vary the relative size of compartments of the level sets $\gamma = 1.0$ versus $\gamma = 2.0$. As an indicator of the effectiveness of CRS we take the number of times CRS(N, m, α) converges to both global minimum points after executing it hundred times. CRS was run for $\alpha = 0.0$ and $\alpha = 0.5$, $m = 2$, 3 and 4 and various values of N.

For $m = 3$, CRS never converges to both minimum points; when $\gamma > 1.0$ it converges to the origin (where the widest compartment is located) and when $\gamma = 1.0$ it randomly converged to one of the global minimum points. Similar results were obtained whenever m was odd, which confirms our analysis from Subsection 3.1.

Table 1. Number of times both optima are detected for the bispherical function with $\gamma = 1.0$ and 2.0

	$\gamma = 1.0$		$\gamma = 2.0$	
	$m = 2$	$m = 4$	$m = 2$	$m = 4$
$N = 100, \alpha = 0.0$	100	100	100	22
$N = 250, \alpha = 0.0$	100	100	100	24
$N = 100, \alpha = 0.5$	70	100	39	0
$N = 250, \alpha = 0.5$	98	100	87	0

Table 2. Number of times both minimum points are detected for the bi-elliptical function ($a = 0.3$, $b = -0.3$, $\phi = \pi/6$, $L_0 = L_\phi = 10.0$ with $\gamma = 1.0$ and $\gamma = 2.0$)

	$\gamma = 1.0$		$\gamma = 2.0$	
	$m = 2$	$m = 4$	$m = 2$	$m = 4$
$N = 100, \alpha = 0.0$	100	100	96	0
$N = 250, \alpha = 0.0$	100	100	100	0
$N = 100, \alpha = 0.5$	25	89	5	0
$N = 250, \alpha = 0.5$	79	100	27	0

For m even, the results in Table 1 show that when $\alpha = 0.5$ (including secondary points) there is more tendency to converge to only one minimum point. Moreover, when the sizes of the compartments differ (γ increases) the effectiveness of CRS can only be ensured when $m = 2$ or when a big population size N is applied.

The geometric component, which stimulates the convergence to both minimum points, is less when the compartments are ellipses which are oriented differently. The consequence is presented in Table 2; the convergence of CRS to both global minimum points is only ensured for $m = 2$ and N big enough. Also from the experiments with bi-elliptical functions we observed that CRS converges to the minimum point corresponding to the compartment (ellipsoid) with the largest volume.

4. Effectiveness in Detecting One Global Minimum Point

Will the algorithm converge to a global minimum point? For this, it is sufficient to consider cases with one global minimum point. There are two possibilities: either the problem has one global and no local optima, or the problem has at least one local, non-global optimum. Leaving out details on the termination criteria, one can say that when there is one minimum point, then it is usually found, based on the argument that CRS can be seen as an extension of the robust local search method of [7]. Counter examples can be constructed by defining a special function f which describes a plateau with a small hole in it containing the minimum. Movement of the population is only possible when strict improvements, $f(\overline{P}) < y_k$, are found, so in such an extreme case no convergence takes place towards the minimum.

The more interesting question is what happens when there are local non-global minimum points. Can the algorithm get trapped in a local non-global minimum? Yes, it can. What can be done about it? To discuss this, let us first introduce some notation. Let $f^* = f(x^*)$ be the minimum value corresponding to (for the ease of the discussion) one global minimum point and f_l^* be the value of the second lowest minimum. During the course of the algorithm, the population spreads over the compartments of the lower level sets corresponding to the local and global minima. Now we can say that the global minimum will be found with certainty at the moment that there are $m+1$ points in $S(f_l^*)$. Those points are never replaced and have a positive probability to be used as parents for generating new primary or secondary points. The probability that $m+1$ points appear in $S(f_l^*)$ is always bigger than zero (and grows with N). Traditional analysis on stochastic methods would tell us that the minimum is found with certainty when we repeat the algorithm infinitely many times.

Let us notice first that this is one reason to choose a big population size N; the probability that $m+1$ points hit $S(f_l^*)$ is bigger when N is bigger and m is smaller. However, the idea of having $m+1$ points in $S(f_l^*)$ is not the sharpest statement which can be made. As explained in the former subsection, when m is even, the geometric component may cause points to be reflected from other compartments of the level set. The sharpest idea is that $m/2$ points in $S(f_l^*)$ are sufficient in the long run to have new points generated in the compartment of the current level set $S(y_k)$ around x^* such that also $m+1$ points are reached, which can serve as parents for primary and secondary points again. In the experiments we observed that it may take a long time, but in limit, independent of the shapes of the other compartments the algorithm converges to x^* when there are at least $m/2$ points in $S(f_l^*)$.

During the course of the algorithms, the compartment of $S(y_k)$ around x^* is desolately far away from the other compartments. When m is odd and there are not $m+1$ points around x^*, or when m is even and there are not $m/2$ points, the algorithm does not converge to the global minimum. An interesting side effect is that the algorithm may not converge at all with respect to traditional stopping criteria on the distance between maximum and minimum function value in the population and size of the final cluster. It may be that $1, 2, \ldots, m$ (for m odd) points are located in $S(f_l^*)$. This means that those points are never thrown away, but also do not reproduce new points in $S(f_l^*)$.

The effectiveness depends on the relative size of the compartment which contains the global minimum point with respect to the size of the total level set. In limit this size is one when $S(f_l^*)$ is reached. One can design experiments to test the effectiveness experimentally. A small example with one local non-global minimum point x_l^* is sufficient to illustrate what happens during the final course of the algorithm. Parameters N, m, and α can be varied and the characteristic of the optimization problem is the relative size of the compartment around the global minimum point. In smooth optimization this is defined by the eigenvalues of the Hessean in x^* compared to the steepness (Hessean) around the local non-global minimum point x_l^*. We design a simple case which is not everywhere smooth, but of which the steepness around the optima can be varied.

$$f(x) = \min \begin{cases} \gamma \cdot [(x_1 - 0.5)^2 + x_2^2], \\ (x_1 + 0.5)^2 + x_2^2 + \rho. \end{cases} \tag{7}$$

The steepness around the global minimum point is determined by the second derivative $2 \times \gamma$. For our experiments we have selected two extreme cases based on the function described by Equation (7), setting $\rho = 0.1$ and for two different values of γ (100.0 and 500.0). The function is constructed such that during the course of the algorithm the size of the compartment around the local minimum is for some time bigger than that of the global minimum. It can be derived that as long as the obtained function value $y > \rho \times \gamma/(\gamma - 1)$ the compartment around the local optimum has a bigger volume than the one around the global minimum point. At the top of Figure 5, a one-dimensional graph for the function is depicted. The figure also illustrates the various courses of the algorithm depicting the location of the individuals as CRS evolves (from left to right). The algorithm results in three possible outcomes:

1. the algorithm finds the global optimum $x^* = (0.5, 0.0)$; the final level set consists of only one compartment where the final population R is located, all the N individuals are clustered around the global minimum point; i.e. $\max f(r_i) \leqslant \varepsilon < \rho, \forall r_i \in R$, where $\varepsilon > 0$ is a tolerance in the stopping conditions.

2. the algorithm converges to the local optimum $x_l^* = (-0.5, 0.0)$; the final level set consists of only one compartment where all the N individuals are located. In this case $\rho \leqslant \min f(r_i) \leqslant \max f(r_i) < \rho + \varepsilon, \forall r_i \in R$.

3. the algorithm does not converge in the sense that less than $m + 1$ (but more than one) points are in $S(f_l^*) = S(\rho)$ for m odd or less than $m/2$ for m even; the final level set consists of two separate compartments. The stopping condition are never met and CRS finishes when a limit on the number of iterations has been reached.

As a performance indicator we define the *Hitting Rate* HR as the number of times the algorithm finds the global minimum point over repeated runs of the algorithm using different (pseudo) random series. In Table 3, experimental results from series of hundred executions of the CRS algorithm for the bispherical problem (7) are given. It should be noticed that for $m = 3$, CRS reached a solution close to the global minimum twelve and two times over hundred, for

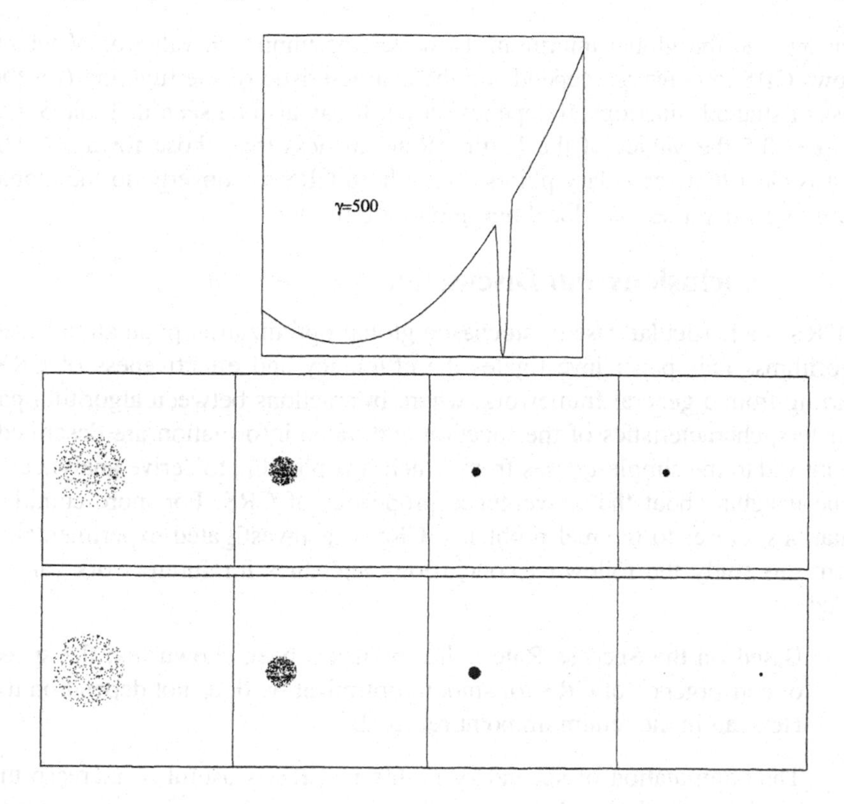

Figure 5. The bispherical problem for $\rho = 0.1$ and $\gamma = 500.0$ (graph at the top) and a graphical representation of evolution of the population points (confined at various level sets) for a real execution of the CRS algorithm with $N = 250$, $\alpha = 0.0$ and $m = 2$ (bottom graphs, first row CRS did not find the global minimum, second row CRS finds the global minimum).

Table 3. Experimental values of the Hitting Rate for bispherical problem (7) with $\rho = 0.1$ and $\gamma = 100.0$ and $\gamma = 500.0$. $^{(*)}$ means that the algorithm did not converge in the sense of case 3.

	$\gamma = 100.0$			$\gamma = 500.0$		
	$m = 2$	$m = 3$	$m = 4$	$m = 2$	$m = 3$	$m = 4$
$N = 100, \alpha = 0.0$	0.31	$0.12^{(*)}$	0.07	0.16	$0.02^{(*)}$	0.00
$N = 250, \alpha = 0.0$	0.64			0.16		
$N = 100, \alpha = 0.5$	0.22			0.12		

$\gamma = 100.0$ and $\gamma = 500.0$, respectively. However, strictly speaking, the algorithm did not converge because only one (or two) individuals over hundred ($N = 100$) were located at $S(\rho)$; i.e. the algorithm did not converge in the sense of case 3 or did not converge to the global minimum at all (case 2).

For m even, under certain conditions, determined by the values of the parameter settings (α and N), CRS is able to converge to the global minimum in the sense of case 1. Table 3 shows that for $m = 2$, if N is great enough CRS converges to the global minimum. However the minimum value of N which allows CRS to converge depends on the characteristic of the function (for the class of studied functions, it depends on γ). It can also be seen in Table 3 that for $\alpha = 0.5$ the values of the Hitting Rate are less than those for $\alpha = 0.0$, so it is clear that secondary points do not help CRS to converge to the global minimum when there is a local non-global minimum.

5. Conclusions and Discussion

CRS is a particular case of stochastic global optimization population based algorithms. This paper investigates the efficiency and effectiveness of CRS. Starting from a general framework, where interactions between algorithm parameters, characteristics of the function and value information are described, we moved to the simplest cases from which it is possible to derive analytically some insights about the convergence properties of CRS. For more complex situations, closer to the real problems, CRS was investigated experimentally. From this study the following conclusions and lines for future work can be posed:

- Based on the Success Rate indicator it has been shown that the speed of convergence of CRS for smooth optimization does not depend on the Hessean in the minimum point reached.
- The computation of secondary points in CRS is useful to increase the efficiency in the final stages of the algorithm when it is converging to one minimum point.
- Secondary points reduce the effectiveness of reaching several global minimum points.
- Convergence to several global minimum points simultaneously is only possible for even values of m. The possibility depends on the Hessean in the global minimum points.
- Values of m greater than 2 do not help the effectiveness of CRS in reaching one or several global minimum points.
- In some specific cases the algorithm may not converge at all.

Bibliography

[1] Hendrix, E. M. T., Ortigosa, P. M. and García, I.: On Success Rates for controlled random search, Technical Note 00-01, Department of Mathematics Wageningen, 2000. (To appear in *JOGO*).

[2] Price, W. L.: A controlled random search procedure for global optimization, In: L. C. W. Dixon and G. P. Szegö (eds), *Towards Global Optimization* 2, North-Holland, Amsterdam, 1978, pp. 71–81.

[3] Price, W. L.: A controlled random search procedure for global optimization, *Comput. J.* **20** (1979), 367–370.

[4] Price, W. L.: Global optimization algorithms by controlled random search, *J. Optim. Theory Appl.* **40** (1983), 333–348.

[5] Klepper, O. and Rouse, D. I.: A procedure to reduce parameter uncertainty for complex models by comparison with real system output illustrated on a potato growth model, *Agricultural Systems* **36** (1991), 375–395.

[6] García, I., Ortigosa, P. M., Casado, L. G., Herman, G. T. and Matej, S.: Multidimensional optimization in image reconstruction from projections, In: I. M. Bomze, T. Csendes, R. Horst and P. M. Pardalos (eds), *Developments in Global Optimization*, Kluwer Acad. Publ., 1997, pp. 289–300.

[7] Nelder, J. A. and Mead, R.: A Simplex method for function minimization, *Comput. J.* **8** (1965), 308–313.

[8] Klepper, O. and Hendrix, E. M. T.: A method for robust calibration of ecological models under different types of uncertainty, *Ecological Modelling* **74** (1994), 161–182.

[9] Hendrix, E. M. T. and Klepper, O.: On uniform covering, adaptive random search and raspberries, *J. Global Optim.* **18** (2000) 143–163.

[10] Törn, A. and Zilinskas, A.: *Global Optimization*, Lecture Notes in Comput. Sci. 350, Springer-Verlag, Berlin, 1989.

[11] Zabinsky, Z. B. and Smith, R. L.: Pure adaptive search in global optimization, *Math. Programming* **53** (1992), 323–338.

[12] Baritompa, W. P., Mladineo, R. H., Wood, G. R., Zabinsky, Z. B. and Baoping, Z.: Towards pure adaptive search, *J. Global Optim.* **7** (1995), 73–110.

Chapter 9

DISCRETE BACKTRACKING ADAPTIVE SEARCH FOR GLOBAL OPTIMIZATION*

Birna P. Kristinsdottir
Mechanical and Industrial Engineering Department
University of Iceland
Hjardarhaga 4
107 Reykjavik
Iceland

Zelda B. Zabinsky
Industrial Engineering, Box 352650
University of Washington
Seattle, WA 98195
USA

Graham R. Wood
Institute of Information Sciences and Technology
Massey University
Palmerston North
New Zealand

Abstract This paper analyses a random search algorithm for global optimization that allows acceptance of non-improving points with a certain probability. The algorithm is called discrete backtracking adaptive search. We derive upper and lower bounds on the expected number of iterations for the random search algorithm to first sample the global optimum. The bounds are derived by modeling the algorithm using a series of absorbing Markov chains. Finally, upper and lower bounds for the expected number of iterations to find the global optimum are derived for specific forms of the algorithm.

Keywords: Global optimization, adaptive search, simulated annealing, random search

*The work of these authors has been supported in part by NSF grant DMI-9820878 and the Marsden Fund administered by the Royal Society of New Zealand.

G. Dzemyda et al. (eds.), Stochastic and Global Optimization, 147–174.
© 2002 *Kluwer Academic Publishers.*

1. Introduction

Random search algorithms are widely used in many areas of science and engineering. Simulated annealing (SA) [1], for example, is a random search algorithm that has been popular for solving both continuous and discrete global optimization problems. The ideas that form the basis of simulated annealing were first published by Metropolis *et al.* in [2] in 1953 and are related to the annealing process of solid materials. In this paper we focus on an algorithm that we call discrete backtracking adaptive search (BAS), a random search algorithm that has a constant probability of accepting a non-improving point at each iteration. By analyzing BAS we hope to gain better insight into how and when accepting non-improving points helps in finding the global optimum, thereby providing a better understanding of simulated annealing type algorithms.

The specific goal of this research is to understand what factors and their interactions affect the expected number of iterations to find the global optimum for a random search algorithm. The three factors that primarily affect the convergence of a random search algorithm are the generator, the acceptance probability and the landscape of the function being optimized. We develop a framework for analyzing random search algorithms by modeling them using absorbing Markov chains.

Complexity of random search algorithms has been investigated by many researchers. In [3] an algorithm that is a combination of pure adaptive search and pure random search was analyzed; the main result showed how the probability of generating a point in the improving region influences the convergence of the algorithm. The effect of including an acceptance probability in the algorithm was not considered. In [4] complexity results were derived for an algorithm called adaptive search that is studied with the purpose of gaining understanding of simulated annealing type algorithms for continuous global optimiation problems. In [5] a random search algorithm called hesitant adaptive search (HAS) that generates iterates in the improving region and also allows hesitation at a current iterate was analyzed, and an expression for the expected number of iterations to find the optimum was derived. In [6] the HAS algorithm was further analyzed and the distribution of the number of iterations to find the optimum was described. Here we extend the HAS analysis by adding a probability of accepting non-improving points for discrete global optimization problems. Convergence properties of simulated annealing (SA) were analyzed in [7] for combinatorial optimization problems, using Markov chains to analyze the algorithm. Among many interesting results it is shown that under certain conditions simulated annealing asymptotically converges to the set of global optimal solutions with probability one. Asymptotic results do not provide information about the expected number of iterations to first find the global optimum, the main focus of our paper. Another application of using Markov chains to study optimization algorithms can be found in [8] where the convergence properties of a canonical genetic algorithm are studied by using homogeneous finite Markov chains.

The paper is organized as follows. Section 2 defines the BAS algorithm and describes its relation to simulated annealing. In Section 3 we derive upper and lower bounds for the expected number of BAS iterations to find an optimum. We also introduce two ways of analyzing the performance of a random search algorithm; the first way is to model the algorithm on the problem domain, and the second way is to model the algorithm on the problem range. We also discuss how these two approaches are related by using the idea of "lumpability". In Section 4 we analyze specific algorithms using our Markov chain models to gain insight into the effect of the acceptance probability on the expected number of iterations to find the global optimum. We look at specific generators for the algorithm and we provide upper and lower bounds on the expected number of iterations for those cases.

2. Discrete BAS and Simulated Annealing

In this paper we consider the following finite global optimization problem:

$$\begin{aligned} &\text{minimize} \quad f(x) \\ &\text{subject to} \quad x \in S, \end{aligned} \tag{1}$$

where f is a real valued function on a finite set S. Consider a sequential random search algorithm that allows acceptance of non-improving points. The algorithm can be characterized by a generator and an acceptance criterion.

To describe the algorithm on the problem domain, we let $x_1, x_2, \ldots, x_K$ denote the K points in the domain S. The generator is described by a probability distribution; d_{ij} is the probability that point x_j is generated from the current point x_i. Similarly the acceptance criterion is described by an acceptance probability t_{ij}, which is the probability that the candidate point x_j is accepted given the current point x_i.

This discrete BAS algorithm can be stated as follows:

DISCRETE BAS.

Step 0. Set $k = 0$ and generate $X_0 \in S$ according to a pre-specified generator.

Step 1. Generate a candidate point $Z \in S$ according to a specific generator, characterized by probability d_{ij}.

Step 2. Set

$$X_{k+1} = \begin{cases} Z & \text{if } f(Z) \leqslant f(X_k), \\ Z & \text{with probability } t_{ij} \text{ if } f(Z) > f(X_k), \\ X_k & \text{otherwise.} \end{cases}$$

Step 3. If a stopping criterion is met, stop. Otherwise increment k and return to Step 1.

There are many ways of generating candidate points in the feasible region S in Step 1. As indicated in Step 1, d_{ij} specifies the probability of generating a candidate point, x_j, from the current point x_i. Discrete BAS is able to capture many types of generators, including neighborhood specific generators. In Section 4 two specific generators will be analyzed for this algorithm. In Step 2, the candidate point is always accepted if it is improving or equal in objective function value to the current point. If the candidate point is worse in value, it is accepted with an acceptance probability. The notation for acceptance probability is similar to the generating probability; t_{ij} specifies the probability of accepting the candidate point, x_j, from the current point, x_i. The acceptance probability can reflect the difference in objective function values between the current point and the candidate point. For instance, if the candidate point is a lot worse than the current point, the acceptance probability may be low, while a candidate point of a slightly worse value may have a relatively high acceptance probability.

The main difference between BAS and simulated annealing is that for BAS the probability of accepting a non-improving point is constant for all iterations of the algorithm, whereas in simulated annealing the acceptance probability changes according to a cooling schedule. That is, instead of t_{ij} being constant as in the BAS algorithm, there is a probability P of accepting a candidate point Z given the current iteration point X_k and the current temperature T_k. This acceptance probability P is typically given by,

$$P(X_k, Z) = \begin{cases} 1 & \text{if } f(Z) \leqslant f(X_k), \\ \exp\left[\dfrac{f(X_k) - f(Z)}{T_k}\right] & \text{otherwise.} \end{cases}$$

This acceptance criteria based on P is also known as the Metropolis criterion [2]. The temperature is controlled by a cooling schedule that gradually reduces the temperature. Initially, when the temperature is high, many non-improving points will be accepted, but as the temperature approaches zero, mostly improving points will be accepted. This feature allows the algorithm to escape from local optima when the temperature is high, but allows convergence to the global optimum when the temperature is low. In the analysis presented here, it is assumed that the temperature is constant, and the effects of a cooling schedule are not considered.

Figure 1 illustrates how the BAS algorithm progresses. The algorithm picks a series of improving points and accepts a non-improving point with a certain probability. We refer to the result of generating and accepting improving points as a "curve". In Figure 1 there are four curves.

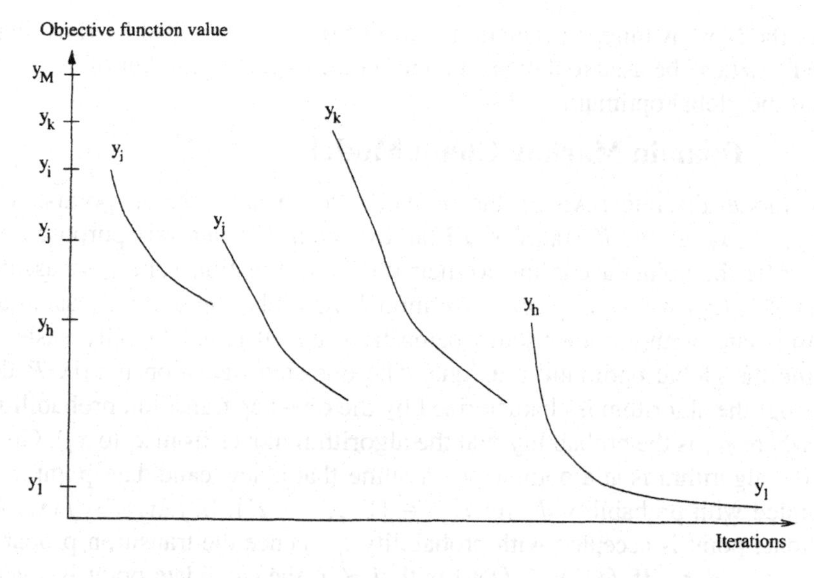

Figure 1. Series of improving points showing acceptance of non-improving points. Each downward run, together with the first higher value is termed a "curve".

3. Modeling Discrete BAS

There are three factors that primarily affect the expected number of iterations to find the optimum: The generator (the mechanism by which iterates are generated in the set S), the acceptance probability (the rule governing acceptance of non-improving points) and the problem structure (which includes number of local minima, convexity and so on). There is an interaction between these factors as they influence the effectiveness of an algorithm to find the global optimum.

To analyze the effects of these factors and their interactions on the expected number of iterations to find the optimum, we describe the discrete BAS random search algorithm using a Markov chain. A *transition* in the Markov chain will correspond to an *iteration* in the random search algorithm. The Markov chain description can be used to track the algorithm in the domain, or in the range. The "domain formulation" is more general than a "range formulation" because it completely describes the algorithm in the domain of a problem. The "range formulation" although less general is more simple to use than the domain formulation. As we will demonstrate, however, not all algorithms can be described exactly using a range formulation. We will show that if the domain formulation satisfies conditions of lumpability [9] it can be reduced to an equivalent Markov range formulation for analysis purposes. For problems that do not satisfy this condition, the domain formulation must be used for a full analysis.

For the Markov range formulation, we develop an "embedded Markov chain model" that can be used to derive a bound on the expected number of iterations to find the global optimum.

3.1. Domain Markov Chain Model

To model discrete BAS on the domain, we consider the K points in S, $x_1, x_2, \ldots, x_K$ as the K states of a Markov chain. For analysis purposes, we now order the points according to their objective function values, so assume $f(x_1) \leqslant f(x_2) \leqslant \cdots \leqslant f(x_K)$. We model the global optimum x_1 as an absorbing state, with all other states being transient. It is notationally easier to assume the global optimum is unique. The one-step transition matrix P that describes the algorithm is characterized by the one-step transition probabilities p_{ij} (where p_{ij} is the probability that the algorithm moves from x_i to x_j). Given that the algorithm is at a point x_i we assume that a new candidate point x_j is generated with probability d_{ij} for $i, j \in \{1, 2, \ldots, K\}$. If $f(x_j) > f(x_i)$ the candidate point is accepted with probability t_{ij}, hence the transition probability is $p_{ij} = d_{ij}t_{ij}$. If $f(x_j) \leqslant f(x_i)$ with $j \neq i$, the candidate point is always accepted, hence the transition probability is $p_{ij} = d_{ij}$. The algorithm stays at the same point if either it is generated again, or another point is generated and rejected. Figure 2 shows the transition matrix P for the domain Markov chain.

The expected number of iterations to first sample the global optimum is equal to the expected number of iterations to absorption and can be expressed in terms of the transition matrix of the Markov chain [9]. We let N be a vector of N_i's where N_i denotes the expected number of iterations to absorption in state x_1, given that the initial starting state was x_i, where $i = 2, \ldots, K$. This expected number of iterations to absorption indicates the average computational effort to first sample the global optimum, but not necessarily to confirm it. The following theorem states how the expected number of iterations to absorption can be found. The proof is in [9].

$$
\begin{array}{c}
\\ K \\ \vdots \\ i \\ \vdots \\ 1
\end{array}
\begin{array}{c}
\begin{array}{ccccc} K & & i & & 1 \end{array} \\
\begin{bmatrix}
d_{KK} & \cdots & d_{Ki} & \cdots & d_{K1} \\
\vdots & \vdots & \vdots & \vdots & \vdots \\
d_{iK}t_{iK} & \cdots & d_{ii} + \displaystyle\sum_{\substack{k=i+1 \\ \text{and} f(x_k) > f(x_i)}}^{K} d_{ik}(1 - t_{ik}) & \cdots & d_{i1} \\
\vdots & \vdots & \vdots & \vdots & \vdots \\
0 & \cdots & 0 & \cdots & 1
\end{bmatrix}
\end{array}
$$

Figure 2. Entries in the one-step transition matrix for a domain Markov chain.

THEOREM 1. *The expected number of iterations N_i prior to absorption in state x_1, starting in state x_i, $i = 2, \ldots, K$, can be found by solving the following system of equations,*

$$N = (I - Q)^{-1}e, \tag{2}$$

where Q, a $(K-1) \times (K-1)$ matrix, is the transient part of the one-step transition matrix P, I is the $(K-1) \times (K-1)$ identity matrix and e is a $K-1$ vector of ones.

Theorem 1 provides a method to calculate the expected number of iterations to absorption (conditional on a given starting state) using the matrix $(I - Q)^{-1}$, often referred to as the fundamental matrix. The variance of the number of iterations until absorption is also obtainable from the fundamental matrix (see [9]).

The domain Markov chain formulation is very general and can be used to model a variety of random search algorithms on various problem domains. However the $I - Q$ matrix may be extremely large, with no apparent structure to aid the inversion. This leads to the development of our second approach, the range Markov chain model.

3.2. Range Markov Chain Model

To facilitate analysis we now develop a Markov chain model of the algorithm on the range of the function, rather than on the problem domain. That is, we let the distinct objective function values $y_1, \ldots, y_M$ denote the states in the Markov chain where $y_1 < y_2 < \cdots < y_M$. Now d_{ij} is the probability of generating a point with objective function value y_j in the problem range, from a point with objective function value y_i. If y_j is less than y_i, the point is always accepted, but if the value y_j is greater than y_i, the non-improving point is accepted with probability t_{ij}. This gives the transition matrix shown in Figure 3. Notice that the number of states in the range is at most the number of states in the domain because more than one point in the problem domain may have the same objective function value. The expected number of iterations for the range Markov chain model can still be found by using the result in Theorem 1.

Using the range Markov chain model provides some benefits to the analysis, as well as putting some restrictions on the type of generator and function that can be analyzed. The primary benefit is being able to solve for the expected number of iterations to absorption more easily, thanks to a smaller and better structured transition matrix, as will be seen in the following subsection. The main restriction is on the transition probabilities. Kemeny and Snell [9] compare an original Markov chain with a reduced state Markov chain called the *lumped Markov chain*, and provide necessary and sufficient conditions for the lumped Markov chain to be equivalent to the original Markov chain with respect to coarser properties, including the expected number of iterations to absorption.

$$
\begin{array}{c} \\ M \\ \vdots \\ i \\ \vdots \\ 1 \end{array}
\begin{array}{c}
\begin{array}{ccccc} M & & i & & 1 \end{array} \\
\begin{bmatrix}
d_{MM} & \cdots & d_{Mi} & \cdots & d_{M1} \\
\vdots & \vdots & \vdots & \vdots & \vdots \\
d_{iM}t_{iM} & \cdots & d_{ii} + \sum_{k=i+1}^{M} d_{ik}(1 - t_{ik}) & \cdots & d_{i1} \\
\vdots & \vdots & \vdots & \vdots & \vdots \\
0 & \cdots & 0 & \cdots & 1
\end{bmatrix}
\end{array}
$$

Figure 3. Entries in the one-step transition matrix for a range Markov chain.

We will reduce the number of states by *lumping* domain states by their range values. Thus we create a partition of the domain Markov chain with states $x_1, \ldots, x_K$, by determining the set of inverse images $f^{-1}(y_j)$ of states for the range Markov chain. We now loosely use y_j to denote the partition set of domain points with objective function value y_j, and the range Markov chain has states $y_1, \ldots, y_M$.

The probability of making a transition from an original domain state x_i to a partition set y_j can be expressed as $\tilde{p}_{x_i y_j} = \sum_{x_k \in y_j} p_{x_i x_k}$, where a domain state x_k is in partition set y_j if $f(x_k) = y_j$. The domain Markov chain is said to be lumpable with respect to the partition if for every pair of partition sets y_i and y_j, the probability $\tilde{p}_{x_l y_j}$ has the same value for every state x_l corresponding to y_i. We denote these common values $\hat{p}_{ij} = \tilde{p}_{x_l y_j}$ for any $x_l \in y_i$. These define the transition probabilities for the *lumped* domain Markov chain. For a given algorithm, the associated range process (the sequence of objective function values) is then Markov precisely when the domain process is lumpable with respect to the partition. In the sequel we shall assume that the range process is Markov.

To illustrate the idea of lumpability, consider an algorithm and a problem that is modeled by the domain Markov chain model. The transition matrix for this problem is shown in Figure 4. In this example there are seven points in the domain, two with objective function value four, two with objective function value three, two with objective function value two and one global optimum with objective function value one. Parameter t is the probability of accepting any non-improving point. This Markov chain is lumpable on the problem domain, giving the range Markov chain shown in Figure 5. Notice that the necessary and sufficient conditions for lumpability for any pair of partition sets y_i and y_j are satisfied. For instance, $\tilde{p}_{x_{4*} y_4} = \sum_{x_k \in y_4} p_{x_{4*} x_k} = p_{x_{4*} x_4} + p_{x_{4*} x_{4*}} = 0.07 + 0.03 = 0.1$, while $\tilde{p}_{x_4 y_4} = \sum_{x_k \in y_4} p_{x_4 x_k} = p_{x_4 x_4} + p_{x_4 x_{4*}} = 0.03 + 0.07 = 0.1$. The reader can verify that the conditions of lumpability are satisfied for other states in the domain Markov chain. These two models

$$\begin{array}{c} \\ 4 \\ 4^* \\ 3 \\ 3^* \\ 2 \\ 2^* \\ 1 \end{array} \begin{array}{c} \begin{array}{ccccccc} 4 & 4^* & 3 & 3^* & 2 & 2^* & 1 \end{array} \\ \left[\begin{array}{ccccccc} 0.03 & 0.07 & 0.05 & 0.15 & 0.1 & 0.2 & 0.4 \\ 0.07 & 0.03 & 0.15 & 0.05 & 0.2 & 0.1 & 0.4 \\ 0.1t & 0.15t & 0.35-0.25t & 0.15 & 0.1 & 0.15 & 0.25 \\ 0.15t & 0.1t & 0.15 & 0.35-0.25t & 0.15 & 0.1 & 0.25 \\ 0.1t & 0.3t & 0.1t & 0.2t & 0.75-0.7t & 0.15 & 0.1 \\ 0.3t & 0.1t & 0.2t & 0.1t & 0.15 & 0.75-0.7t & 0.1 \\ 0 & 0 & 0 & 0 & 0 & 0 & 1 \end{array}\right] \end{array}$$

Figure 4. Domain Markov chain model for lumpability example.

$$\begin{array}{c} \\ 4 \\ 3 \\ 2 \\ 1 \end{array} \begin{array}{c} \begin{array}{cccc} 4 & 3 & 2 & 1 \end{array} \\ \left[\begin{array}{cccc} 0.1 & 0.2 & 0.3 & 0.4 \\ 0.25t & 0.25+(1-t)0.25 & 0.25 & 0.25 \\ 0.4t & 0.3t & 0.2+(1-t)0.7 & 0.1 \\ 0 & 0 & 0 & 1 \end{array}\right] \end{array}$$

Figure 5. Range Markov chain model for lumpability example.

are equivalent in terms of the distribution of the number of iterations to absorption, and hence the domain and range processes have the same expected number of iterations to first find the global optimum.

3.3. Range Embedded Markov Chain Model

If only improving points are accepted by an algorithm, the resulting range transition matrix is upper triangular so can in many instances be inverted analytically. If acceptance of non-improving points is allowed, the range $I - Q$ matrix is full and therefore difficult to invert analytically. To derive upper and lower bounds for the expected number of iterations to find the optimum, we model the algorithm as a series of embedded Markov chains.

Referring back to Figure 1, the optimization algorithm produces a series of improving points that are always accepted, and then occassionally accepts a non-improving point. This process can be modeled as a series of Markov chains, where one embedded Markov chain corresponds to a curve of improving points. The embedded Markov chain transition matrix is then upper triangular because it only includes improving points. The accepted non-improving point then terminates one curve and starts another.

To model the algorithm with a series of embedded Markov chains, the M states of the Markov chain represent the objective function values, $y_1, \ldots, y_M$. The global optimum y_1 is an absorbing state and will be denoted 1, while the transient states include $2, \ldots, M$. To model acceptance of non-improving points, added to the model are distinct absorbing states, $2', \ldots, M'$ that reflect the event of accepting a non-improving point. That is, when a non-improving point j is accepted the chain reaches an absorbing state j'. A "curve" is formed when the chain reaches an absorbing state.

To describe the associated one-step transition matrix of the embedded Markov chain, illustrated in Figures 6 and 7, suppose the process is in a transient state i, where $i \in \{2, \ldots, M\}$. Three types of points could be sampled. First, the global optimum could be sampled, in which case the Markov chain is absorbed into state 1. Because this is always an improving point, the transition probability is just the probability that the global optimum is sampled, $p_{i1} = d_{i1}$, for $i \in \{2, \ldots, M\}$. Second, an intermediate non-absorbing improving point could be sampled, and in this case it is always accepted, so $p_{ij} = d_{ij}$ for $1 < j < i$ and $i \in \{2, \ldots, M\}$. Third, a non-improving point y_j could be sampled, with probability d_{ij}, for $j > i$. This point is either accepted, with probability t_{ij}, or rejected. If it is accepted, the Markov chain moves to absorbing state j', and the transition probability is $p_{ij'} = d_{ij}t_{ij}$. If the point is rejected, the Markov chain stays at its current point i. Thus the transition probability from state i to i, for $i \in \{2, \ldots, M\}$, is the chance it is sampled directly plus the sum of probabilities that higher points are sampled and rejected, or $p_{ii} = d_{ii} + \sum_{k=i+1}^{M} d_{ik}(1 - t_{ik})$.

Figures 6 and 7 illustrate the one-step transition matrix P. In Figure 6, the structure of the one-step transition matrix is shown in blocks, where Q^* includes transitions to the same or improving points and R^* includes transitions to non-improving points. Notice that Q^* is an upper triangular matrix and R^* is a lower triangular matrix. Both are $(M - 1) \times (M - 1)$ matrices. Figure 7 illustrates the detailed one-step transition matrix, and for $i \in \{2, \ldots, M\}$, the Q^* and R^* matrices have the following entries:

$$q^*_{ij} = \begin{cases} 0 & \text{if } i < j \text{ and } j \in \{2, \ldots, M\}, \\ d_{ii} + \sum\limits_{k=i+1}^{M} d_{ik}(1 - t_{ik}) & \text{if } i = j \text{ and } j \in \{2, \ldots, M\}, \\ d_{ij} & \text{if } i > j \text{ and } j \in \{2, \ldots, M\}, \end{cases} \tag{3}$$

$$r^*_{ij'} = \begin{cases} d_{ij}t_{ij} & \text{for } j > i \text{ and } j' \in \{2', \ldots, M'\}, \\ 0 & \text{for } j \leqslant i \text{ and } j' \in \{2', \ldots, M'\}. \end{cases} \tag{4}$$

	$M \;\ldots\; 2$	1	$M' \;\ldots\; 2'$
$M \;\vdots\; 2$	Q^*	d_{i1}	R^*
1	0	1	0
$M' \;\vdots\; 2'$	0	0	I

Figure 6. Structure of one-step transition matrix, P.

	M	$\dots$	i	$\dots$	2	1	M'	$\dots$	$(i+1)'$	$\dots$	$2'$
M	d_{MM}	$\dots$	d_{Mi}	$\dots$	d_{M2}	d_{M1}	0	$\dots$	0	$\dots$	0
$\vdots$	$\vdots$	$\vdots$	$\vdots$	$\vdots$	$\vdots$	$\vdots$	$\vdots$	$\vdots$	$\vdots$	$\vdots$	$\vdots$
i	0	$\dots$	$d_{ii} + \sum_{k=i+1}^{M} d_{ik}(1 - t_{ik})$	$\dots$	d_{i2}	d_{i1}	$d_{iM}t_{iM}$	$\dots$	$d_{ii+1}t_{ii+1}$	$\dots$	0
$\vdots$	$\vdots$	$\vdots$	$\vdots$	$\vdots$	$\vdots$	$\vdots$	$\vdots$	$\vdots$	$\vdots$	$\vdots$	$\vdots$
2	0	$\dots$	0	$\dots$	d_{22}	d_{21}	$d_{2M}t_{2M}$	$\dots$	$d_{2i+1}t_{2i+1}$	$\dots$	0
1	0	$\dots$	0	$\dots$	0	1	0	$\dots$	0	$\dots$	0
M'	0	$\dots$	0	$\dots$	0	0	1	$\dots$	0	$\dots$	0
$\vdots$	$\vdots$	$\vdots$	$\vdots$	$\vdots$	$\vdots$	$\vdots$	$\vdots$	$\vdots$	$\vdots$	$\vdots$	$\vdots$
$(i+1)'$	0	$\dots$	0	$\dots$	0	0	0	$\dots$	1	$\dots$	0
$\vdots$	$\vdots$	$\vdots$	$\vdots$	$\vdots$	$\vdots$	$\vdots$	$\vdots$	$\vdots$	$\vdots$	$\vdots$	$\vdots$
$2'$	0	$\dots$	0	$\dots$	0	0	0	$\dots$	0	$\dots$	1

Figure 7. Entries in one-step transition matrix, P.

The following theorem states how it is possible to find the expected number of iterations to find the global optimum when the algorithm is modeled using the embedded Markov chain.

THEOREM 2. *The expected number of iterations N_i to absorption in state* 1, *starting in state i, $i = 2, \ldots, M$, can be found by solving the following system of equations*

$$N = (I - F^*)^{-1} u, \tag{5}$$

where F^* and u can be found by solving the following system of equations,

$$u = (I - Q^*)^{-1} e, \tag{6}$$

$$F^* = (I - Q^*)^{-1} R^*, \tag{7}$$

where Q^* and R^* are defined in Equations (3) *and* (4).

The proof is in the Appendix.

Theorem 2 presents an expression for the expected number of iterations to find the optimum for the embedded Markov chain model by solving three systems of equations. All the equations require the inversion of a matrix. The matrix inversions in (6) and (7) are relatively easy because Q^* is upper triangular, but the matrix $(I - F^*)$ in (5) is a full matrix which makes an analytical inversion extremely difficult. However, it is possible to get an upper and a lower bound for the expected number of iterations N by using (6) and (7). The following theorem gives an upper bound and a lower bound for the expected number of iterations to first find the global optimum.

THEOREM 3. *The expected number of iterations N_i to first sample the global optimum, starting in state i is bounded above by*

$$u_i + \max_{j \in \{2,\ldots,M\}} \left\{ \left(\frac{u_j}{f_{j1}} \right) \right\} (1 - f_{i1}) \tag{8}$$

and is bounded below by

$$u_i + \min_{j \in \{2,\ldots,M\}} \left\{ \left(\frac{u_j}{f_{j1}} \right) \right\} (1 - f_{i1}), \tag{9}$$

where u_j and f_{j1} are as defined in Equations (6) *and* (7).

The proof is in the Appendix.

Notice that the upper and lower bounds are expressed in terms of u_i and f_{i1}. Here u_i is the expected number of iterations to absorption in any absorbing state, given that the algorithm started in state i, and f_{i1} is the probability of being absorbed in state 1 (the global optimum), given that the algorithm started in state i. Also notice that deriving u_i and f_{i1} only requires inversion of an upper triangular matrix.

4. Complexity Analysis of Specific Algorithms

The results presented in Theorems 2 and 3 hold for a Markov range algorithm with generator d_{ij} and acceptance probability t_{ij}. We will now derive analytical and numerical complexity results for algorithms that use specific generators. Complexity is measured in terms of the expected number of iterations to first sample the global optimum. We will look at two extreme algorithms. The first example, Example A, illustrates an algorithm where the probability of sampling the global optimum decreases the closer we are in objective function value to the global optimum. This may reflect a ridge surrounding the global optimum. The second example, Example B, illustrates an algorithm where the probability of sampling the global optimum increases the closer we are in objective function value to the global optimum. This may be realized by a convex function. We analyze how the acceptance probability affects the expected number of iterations to first find the global optimum for both algorithms, and we derive upper and lower bounds for the expected number of iterations for both algorithms. We demonstrate that the exact expected number of iterations can be close to either bound. We also demonstrate that a positive probability of accepting a non-improving point improves performance in Example A, but degrades performance in Example B.

Example A. To illustrate an algorithm where the probability of sampling the global optimum decreases the closer we are in objective function value to the global optimum, consider the example described by the transition matrix shown in Figure 8. This is the example used in Figures 2 and 3.

$$
\begin{array}{c} \\ 4 \\ 3 \\ 2 \\ 1 \end{array}
\begin{array}{c}
\begin{array}{cccc} 4 & 3 & 2 & 1 \end{array} \\
\left[\begin{array}{cccc}
0.1 & 0.2 & 0.3 & 0.4 \\
0.25t & 0.25+(1-t)0.25 & 0.25 & 0.25 \\
0.4t & 0.3t & 0.2+(1-t)0.7 & 0.1 \\
0 & 0 & 0 & 1
\end{array}\right]
\end{array}
$$

Figure 8. Range Markov chain model for Example A.

Notice that at objective function value 4 there is a fairly high probability of sampling the global optimum, 0.4, while at objective function value 3 this probability is lower, 0.25, and even lower at objective function value 2, where it is 0.1. We use a single parameter t for the probability of accepting a non-improving point, so we can explore the effect as t ranges between 0 and 1.

Figure 9 shows the upper and lower bounds on expected number of iterations by using the bounds in Theorem 3 and the exact expected number of iterations

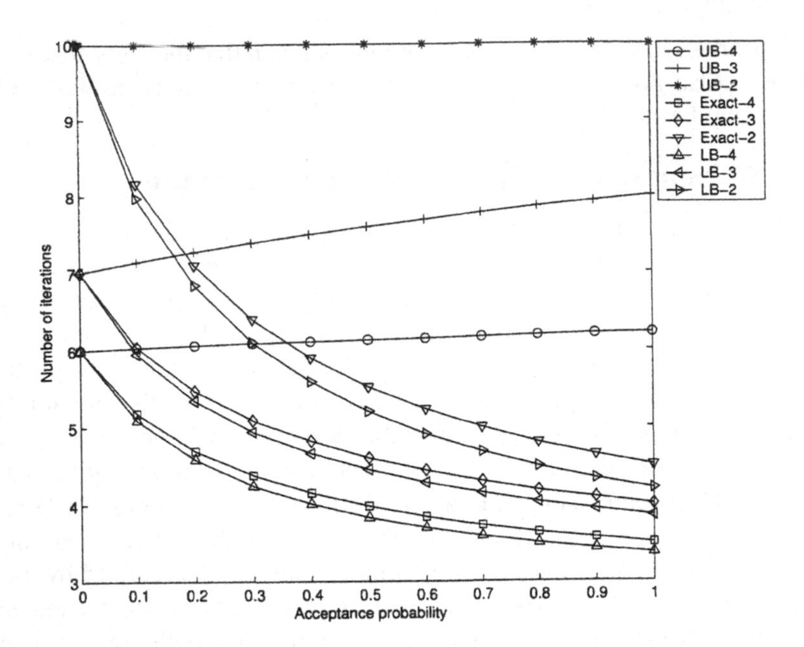

Figure 9. Upper bound, lower bound and exact number of iterations to convergence for Example A, for starting points of 4, 3 or 2.

found with either Theorem 1 or Theorem 2 for starting points 4, 3, or 2. The figure shows that increasing the acceptance probability helps the algorithm find the optimum in fewer iterations. For this type of generator and "landscape" it is beneficial to accept a non-improving point. The second fact to notice is how close the exact expected number of iterations is to the lower bound for the expected number of iterations.

Example B: Combined PAS/PRS Algorithm with Acceptance Probability. In an attempt to provide insight into the performance of general random search algorithms, an algorithm that is a combination of pure adaptive search (PAS) [10,11] and pure random search (PRS) is now examined. This combination of PAS and PRS was first described in [3] by Zabinsky and Kristinsdottir, where the combined algorithm samples according to PAS with probability p, and according to PRS with probability $1-p$, both based on the same generating distribution. In [3] it was demonstrated that even a small value of p can make a large improvement in performance. The analysis in [3] did not allow acceptance of non-improving points (so $t_{ij} = 0$), and here we extend that analysis by using the embedded Markov chain to capture the effects of allowing a positive probability t of accepting a non-improving point. There are two versions of the PAS algorithm for discrete domains, a weak version and a strong version [12]. In this paper we use the weak version of PAS where the level set is $S(X_k) = \{x : x \in S \text{ and } f(x) \leqslant f(X_k)\}$.

Consider the finite global optimization problem that was presented earlier. As before let $y_1 < y_2 < \cdots < y_M$ be distinct objective function values. In keeping with the notation in [12], pure random search samples the domain according to a fixed probability distribution μ on S, independently of previous points. Given this sampling distribution, a probability measure $\pi = (\pi_1, \ldots, \pi_M)$ is defined on the range of f as follows. Let π_j be the probability that any iteration of pure random search attains a value of y_j, that is, $\pi_j = \mu(f^{-1}(y_j))$ for $j = 1, \ldots, M$. Let $p_j = \sum_{i=1}^{j} \pi_i$, the probability that PRS attains a value of y_j or less. Given the current objective function level y_i, the probability that PAS samples y_j, for $j < i$, is π_j / p_i.

The embedded Markov chain model just presented can be used to model this PAS/PRS mixture algorithm that has a non-zero probability of accepting non-improving points. The transition probability will incorporate the probability of sampling according to PAS or PRS as well as the probability of accepting the sampled point even if it is not improving. This one-step transition matrix has the same structure as in Figure 6, with entries in the Q^* and R^* matrices the following, for $i = 2, 3, \ldots, M$,

$$q^*_{ij} = \begin{cases} 0 & i < j, j \in \{2, \ldots, M\}, \\ \displaystyle\sum_{k=i+1}^{M} (1-p)\pi_k(1-t_{ik}) + & \\ \qquad + (1-p)\pi_j + p\left(\dfrac{\pi_j}{p_i}\right) & i = j, j \in \{2, \ldots, M\}, \\ (1-p)\pi_j + p(\pi_j/p_i) & i > j, j \in \{2, \ldots, M\}, \end{cases} \tag{10}$$

$$r^*_{ij'} = \begin{cases} (1-p)\pi_j t_{ij} & \text{if } i < j, j' \in \{2', \ldots, M'\}, \\ 0 & \text{if } i \geqslant j, j' \in \{2', \ldots, M'\}. \end{cases} \tag{11}$$

When j' is strictly larger than i, the only way the algorithm can move from state i to j' is if state j' is sampled according to pure random search, and the point is accepted, thus $p_{ij'} = (1-p)\pi_j t_{ij}$. When $i = j$ the algorithm will stay at i by sampling a non-improving point using PRS and rejecting it, or by sampling the current point again with PAS or PRS. When j is strictly less than i, then state j can be sampled according to PRS or PAS, thus $p_{ij} = (1-p)\pi_j + p(\pi_j/p_i)$, and the improving point is always accepted.

The PAS/PRS algorithm is now analyzed for the special case where a uniform distribution is used for the generator and the acceptance probability t is constant. Assuming a uniform distribution gives $\pi_j = 1/M$ and $p_i = \sum_{k=1}^{i} \pi_k = i/M$. Using these assumptions we get the transition matrix shown in Figure 10.

Theorem 3 presented an analytical upper bound for the expected number of iterations to find the optimum. For our particular combination of generators, PAS and PRS, it is true that $N_M \geqslant N_i$ for $i = 2, \ldots, M$, because for PAS

	M	$\dots$	i	$\dots$	2	1	M'	$\dots$	$(i+1)'$	$\dots$	$2'$
M	$\frac{1-p}{M} + \frac{p}{M}$	$\dots$	$\frac{1-p}{M} + \frac{p}{M}$	$\dots$	$\frac{1-p}{M} + \frac{p}{M}$	$\frac{1-p}{M} + \frac{p}{M}$	0	$\dots$	0	$\dots$	0
$\vdots$	$\vdots$	$\vdots$	$\vdots$	$\vdots$	$\vdots$	$\vdots$	$\vdots$	$\vdots$	$\vdots$	$\vdots$	$\vdots$
i	0	$\dots$	$\frac{1-p}{M} + \frac{p}{i} + \frac{(M-i)(1-t)(1-p)}{M}$	$\dots$	$\frac{1-p}{M} + \frac{p}{i}$	$\frac{1-p}{M} + \frac{p}{i}$	$\frac{(1-p)t}{M}$	$\dots$	$\frac{(1-p)t}{M}$	$\dots$	0
$\vdots$	$\vdots$	$\vdots$	$\vdots$	$\vdots$	$\vdots$	$\vdots$	$\vdots$	$\vdots$	$\vdots$	$\vdots$	$\vdots$
2	0	$\dots$	0	$\dots$	$\frac{1-p}{M} + \frac{p}{2} + \frac{(M-1)(1-t)(1-p)}{M}$	$\frac{1-p}{M} + \frac{p}{2}$	$\frac{(1-p)t}{M}$	$\dots$	$\frac{(1-p)t}{M}$	$\dots$	0
1	0	$\dots$	0	$\dots$	0	1	0	$\dots$	0	$\dots$	0
M'	0	$\dots$	0	$\dots$	0	0	1	$\dots$	0	$\dots$	0
$\vdots$	$\vdots$	$\vdots$	$\vdots$	$\vdots$	$\vdots$	$\vdots$	$\vdots$	$\vdots$	$\vdots$	$\vdots$	$\vdots$
$(i+1)'$	0	$\dots$	0	$\dots$	0	0	0	$\dots$	1	$\dots$	0
$\vdots$	$\vdots$	$\vdots$	$\vdots$	$\vdots$	$\vdots$	$\vdots$	$\vdots$	$\vdots$	$\vdots$	$\vdots$	$\vdots$
$2'$	0	$\dots$	0	$\dots$	0	0	0	$\dots$	0	$\dots$	1

Figure 10. Entries in transition matrix for mix of PAS and PRS, uniform distribution, constant t.

the expected number of iterations to find the optimum is largest when starting at state M and the expected number of iterations for PRS is the same for all starting states. When it is known which state ($i = \max$) gives the maximum number of iterations, that is $N_{\max} = \max\{N_2, \ldots, N_M\}$, the upper bound simplifies to $N_i \leqslant u_i + (u_{\max}/f_{\max,1})(1 - f_{i1})$. Using $M = \max$ we can write the upper bound as $N_M \leqslant u_M/f_{M1}$. Finding this bound does not require an inversion of $I - F^*$, only an inversion of $I - Q^*$, shown in Figure 12 to find u_M and f_{1M}. This can be done analytically because for the embedded Markov chain model $I - Q^*$ is upper triangular and has a special structure. The lower bound can be found similarly, using $N_2 \geqslant u_2/f_{21}$.

The following theorem gives upper and lower bounds for the expected number of iterations N_M to first find the global optimum starting in state M, in terms of p, the probability of generating a point in the improving region using PAS, and t, the probability of accepting a non-improving point. Deriving analytical information for N will give valuable insight into how the acceptance probability t and the probability p affect the expected number of iterations to find the optimum.

THEOREM 4. *The expected number of iterations to solve the global optimization problem for PAS/PRS with probability p, assuming a uniform distribution for the generating method and a constant acceptance probability t, and starting in state M, is bounded above by*

$$\frac{1 + \sum_{j=2}^{M-1} \frac{j}{\rho_j} \prod_{k=j+1}^{M-1} (1 + \frac{\gamma_k}{\rho_k})}{\frac{\gamma_M}{\rho_M}(1 + \sum_{j=2}^{M-1} \frac{\gamma_j}{\rho_j} \prod_{k=j+1}^{M-1} (1 + \frac{\gamma_k}{\rho_k}))}$$

and is bounded below by

$$\frac{2M}{pM + 2(1-p)},$$

where γ_k and ρ_k are given by,

$$\gamma_k = pM + k(1-p),$$

and

$$\rho_k = (k-1)\left(pM + k(1-p) + \left(\frac{k}{k-1}\right)(1-p)(M-k)t\right).$$

The proof follows directly from Theorem 3 where we derived an upper bound for the expected number of iterations as u_M/f_{M1} and a lower bound on the expected number of iterations as u_2/f_{21}. Expressions for u_i and f_{i1} are given in the Appendix.

Figure 11 shows the upper and lower bounds on expected number of iterations of the combined PAS/PRS algorithm compared to the exact expected

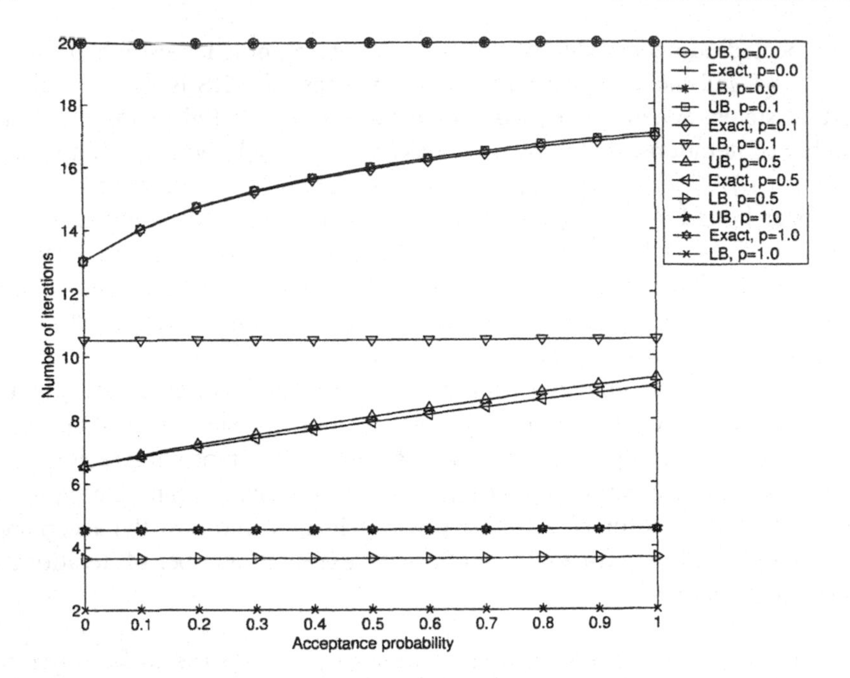

Figure 11. Upper bound (UB), lower bound (LB) and exact expected number of iterations to convergence for Example B for $p = 0.0$, $p = 0.1$, $p = 0.5$ and $p = 1.0$.

number of iteration for four different p values, 0.0, 0.1, 0.5, 1.0 and $M = 20$. Notice when $p = 1.0$ the algorithm is PAS, and when $p = 0.0$ the algorithm is PRS. The exact number of iterations is calculated numerically by solving for N as $(I - Q)^{-1}e$ for 20 states. Notice how close the upper bound is to the exact expected number of iterations. Contrast this with Example A where the exact number of iterations was very close to the lower bound. Also notice how higher acceptance probability does not speed up the algorithm; on the contrary, as the acceptance probability increases the expected number of iterations increases also. That is, adding acceptance probability to the PAS/PRS algorithm seems to only degrade the peformance of the algorithm, while Example A shows just the opposite. In the PAS/PRS algorithm the probability of sampling the global optimum increases or stays the same as the objective function value decreases. Therefore, there is no incentive to accept a non-improving point; such action will only degrade performance because the probability of sampling the global optimum will be lower.

5. Conclusions

This paper presents an analysis of random search algorithms that can accept non-improving points. Upper and lower bounds on the expected number of iterations to first find a global optimum for a random search algorithm are derived. We began with a Markov chain on the domain, and then developed a Markov chain on the range providing we assume the domain Markov chain is lumpable. We then developed an embedded Markov chain model to make analysis possible using upper and lower triangular matrices.

The results provide insight into the relationship between generator, acceptance probability and problem structure for a global optimization problem. Two examples were used to demonstrate the different effects of including an acceptance probability and how this relates to the generator and problem structure. In the first example, we analyzed an algorithm where the probability of sampling the global optimum decreases as we get close in objective function value to the optimum. In this case including acceptance probability improves performance of the algorithm. The exact expression for the expected number of iterations was close to the derived lower bound. In the second example we analyzed an algorithm that is a combination of PAS and PRS, where the probability of sampling the global optimum increases as we get closer in objective function value to the optimum. In this example acceptance probability only distracts the algorithm from sampling points that are helpful in moving closer to the global optimum. For this example the exact expression for the expected number of iterations was close to the derived upper bound.

Appendix

PROOF OF THEOREM 2. Let u_i denote the expected number of iterations until absorption into *any* of the absorbing states, starting in state i. Solving the following system of equations determines u_i (see [13]), for $i = 2, \ldots, M$,

$$u_i = p_{i1} + \sum_{j=2}^{M} r^*_{ij'} + \sum_{k=2}^{M} q^*_{ik}(1 + u_k) \tag{12}$$

$$= 1 + \sum_{k=2}^{M} q^*_{ik} u_k, \tag{13}$$

where q^*_{ik} and $r^*_{ij'}$ are given in Equations (3) and (4). In matrix notation this can be written as

$$u = (I - Q^*)^{-1} e. \tag{14}$$

Note that u gives the expected number of iterations to be absorbed into *any* one of the absorbing states, but not in a specific one.

Let $f_{ij'}$ be the probability of being absorbed into a specific state j', starting in state i, then

$$f_{i1} = p_{i1} + \sum_{k=2}^{M} q^*_{ik} f_{k1}, \tag{15}$$

$$f_{ij'} = r^*_{ij'} + \sum_{k=2}^{M} q^*_{ik} f_{kj'}. \tag{16}$$

In matrix notation this can be written as

$$F = (I - Q^*)^{-1} R.$$

The matrix F is an $(M-1) \times M$ matrix and F^* is the $(M-1) \times (M-1)$ matrix that results when the column corresponding to $j = 1$ is not included. The matrices R^* and R are related in the same way.

Let $m_{ij'}$ denote the conditional mean first passage time, given that a passage to j' is certain to occur. Then $m_{ij'}$ can be found as follows [13],

$$f_{ij'} m_{ij'} = f_{ij'} + \sum_{k=2}^{M} q^*_{ik} f_{kj'} m_{kj'}. \tag{17}$$

When there is only one absorbing state, (17) reduces to (13). If there is more than one absorbing state then $\sum_{k=1}^{M} f_{ik'} m_{ik'} = u_i$. Summing both sides of Equation (17) gives Equation (13),

$$\sum_{j=1}^{M} f_{ij'} m_{ij'} = \sum_{j=1}^{M} f_{ij'} + \sum_{j=1}^{M} \sum_{k=2}^{M} q^*_{ik} f_{kj'} m_{kj'},$$

$$u_i = 1 + \sum_{k=2}^{M} q^*_{ik} \sum_{j=1}^{M} f_{kj'} m_{kj'}$$

$$= 1 + \sum_{k=2}^{M} q^*_{ik} u_k.$$

The following equation can be set up to find N_i, the expected number of iterations to absorption in the global optimum 1, starting in i,

$$N_i = f_{i1} m_{i1} + \sum_{k=2}^{M} f_{ik'} (m_{ik'} + N_k), \tag{18}$$

where $f_{ij'}$ comes from (16) and $m_{ij'}$ comes from (17). The reasoning for (18) is by a first step analysis. Assume the optimization starts in state i then the first curve could end in an absorption in 1 (expected number of iterations $f_{i1} m_{i1}$) or it can end in state k' where $k' \in \{2', \ldots, M'\}$ with a probability $f_{ik'}$, taking $m_{ik'}$ iterations. The next curve will start in state k and therefore it is going to take an additional N_k iterations to absorption in 1 or $1'$ (we are assuming that $1 = 1'$).

Equation (18) can be rewritten as follows,

$$N_i = \sum_{k=1}^{M} f_{ik'} m_{ik'} + \sum_{k=2}^{M} f_{ik'} N_k. \tag{19}$$

Using that $\sum_{k=1}^{M} f_{ik'} m_{ik'} = u_i$, Equation (19) can be rewritten as

$$N_i = u_i + \sum_{k=2}^{M} f_{ik'} N_k \tag{20}$$

or in matrix form as

$$(I - F^*)N = u. \tag{21}$$

□

PROOF OF THEOREM 3. Using equation $u = (I - F^*)N$ we get for u_i

$$\begin{aligned} u_i &= -f_{iM} N_M - f_{i,M-1} N_{M-1} + \cdots + (1 - f_{ii}) N_i + \cdots - f_{i2} N_2 \\ &= N_i - (f_{iM} N_M + \cdots + f_{i2} N_2) \end{aligned}$$

letting $N_{\max} = \max\{N_2, \ldots, N_M\}$, gives

$$\begin{aligned} u_i &\geqslant N_i - N_{\max}(f_{iM} + \cdots + f_{i2}) \\ &\geqslant N_i - N_{\max}(1 - f_{i1}), \quad \text{or} \\ N_i &\leqslant u_i + N_{\max}(1 - f_{i1}). \end{aligned} \tag{22}$$

At $i = \max$, that is, the starting state that gives the largest N_i, we get using Equation (22),

$$\begin{aligned} N_{\max} &\leqslant u_{\max} + N_{\max}(1 - f_{\max,1}), \quad \text{or} \\ N_{\max} &\leqslant \frac{u_{\max}}{f_{\max,1}} \end{aligned}$$

and using $N_{\max} \leqslant \frac{u_{\max}}{f_{\max,1}}$ in Equation (22) gives

$$N_i \leqslant u_i + \frac{u_{\max}}{f_{\max,1}}(1 - f_{i1}).$$

The above equation assumes it is known where N_i attains its maximum. This assumption may not be realistic, so we replace $u_{\max}/f_{\max,1}$ with a number that is greater than or equal to that ratio, $\max\{\frac{u_j}{f_{j1}}\}$ over $j = 1, \ldots, M$. Therefore we get $u_i + \frac{u_{\max}}{f_{\max,1}}(1 - f_{i1}) \leqslant u_i + \max_{j \in \{2,\ldots,M\}}\{\frac{u_j}{f_{j1}}\}(1 - f_{i1})$ and therefore we get the following bound in terms of known quantities,

$$N_i \leqslant u_i + \max_{j \in \{2,\ldots,M\}} \left\{ \left(\frac{u_j}{f_{j1}} \right) \right\} (1 - f_{i1}).$$

The lower bound can be found similarly,

$$\begin{aligned} u_i &= -f_{iM}N_M - f_{i,M-1}N_{M-1} + \cdots + (1 - f_{ii})N_i + \cdots - f_{i2}N_2 \\ &= N_i - (f_{iM}N_M + \cdots + f_{i2}N_2) \end{aligned}$$

and letting $N_{\min} = \min\{N_2, \ldots, N_M\}$, gives

$$\begin{aligned} u_i &\leqslant N_i - N_{\min}(f_{iM} + \cdots + f_{i2}) \\ &\leqslant N_i - N_{\min}(1 - f_{i1}), \quad \text{or} \\ N_i &\geqslant u_i + N_{\min}(1 - f_{i1}). \end{aligned} \tag{23}$$

When $i = \min$, the starting state that gives the smallest N_i, we find using Equation (23),

$$\begin{aligned} N_{\min} &\geqslant u_{\min} + N_{\min}(1 - f_{\min,1}), \quad \text{or} \\ N_{\min} &\geqslant \frac{u_{\min}}{f_{\min,1}} \end{aligned}$$

and using $N_{\min} \geqslant \frac{u_{\min}}{f_{\min,1}}$ in Equation (23) gives

$$N_i \geqslant u_i + \frac{u_{\min}}{f_{\min,1}}(1 - f_{i1}).$$

This assumes it is known where N_i attains its minimum. This assumption may not always be realistic, so we replace $u_{\min}/f_{\min,1}$ with a number that is greater than or equal to that ratio, $\min\{\frac{u_j}{f_{j1}}\}$ over $j = 1, \ldots, M$. The bound can then be rewritten in terms of known quantities as

$$N_i \geqslant u_i + \min_{j\in\{2,\ldots,M\}} \left\{ \left(\frac{u_j}{f_{j1}} \right) \right\} (1 - f_{i1}). \qquad \square$$

THEOREM 5. *The expected number of iterations to absorption in any of the absorbing states for the embedded Markov chain model of PAS/PRS, assuming a uniform distribution for the generating method and starting in state y_i for $i = 2, \ldots, M$ is*

$$u_i = \frac{iM}{\rho_i} + \frac{M\gamma_i}{\rho_i} \left\{ \frac{i-1}{\rho_{i-1}} + \sum_{j=2}^{i-2} \frac{j}{\rho_j} \prod_{k=j+1}^{i-1} \left(1 + \frac{\gamma_k}{\rho_k} \right) \right\}, \tag{24}$$

where p is the probability of sampling according to PAS, t is the acceptance probability and γ_k and ρ_k are defined as,

$$\gamma_k = pM + k(1 - p)$$

and

$$\rho_k = (k - 1) \left(pM + k(1 - p) + \left(\frac{k}{k - 1} \right) (1 - p)(M - k)t \right).$$

For the special case when $i = M$ and $i = 2$ this simplifies to

$$u_M = 1 + \sum_{j=2}^{M} \frac{j}{\rho_j} \prod_{k=j+1}^{M} \left(1 + \frac{\gamma_k}{\rho_k}\right), \tag{25}$$

$$u_2 = \frac{2M}{\rho_2}. \tag{26}$$

PROOF. The transition matrix is upper triangular and therefore back substitution can be used to solve for u_i, for $i = 2, \ldots, M$. The formula for back substitution for this problem is

$$u_i = \frac{1 - \sum_{j=2}^{i-1} a_{ij} u_j}{a_{ii}}, \tag{27}$$

where a_{ij} denotes the entries in the matrix $(I - Q^*)$ (see Figure 12). The matrix in Figure 10 can be derived from the matrix shown in Figure 7. For instance, consider the iith element in the $(I - Q^*)$ matrix,

$$\begin{aligned} 1 - \frac{1-p}{M} - \frac{p}{i} - \frac{(1-p)(M-i)(1-t)}{M} \\ = \frac{Mi - i(1-p) - pM - i(1-p)(M-i)(1-t)}{Mi} \\ = \frac{Mi - i(1-p)M + i(i-1)(1-p) - pM + i(1-p)(M-i)t}{Mi} \\ = \frac{(i-1)(pM + i(1-p) + (\frac{i}{i-1})(1-p)(M-i)t}{Mi} \\ = \frac{\rho_i}{Mi}. \end{aligned}$$

$$\begin{array}{c} \\ M \\ \vdots \\ i \\ \vdots \\ 2 \end{array} \begin{array}{c} \begin{array}{ccccc} M & \cdots & i & \cdots & 2 \end{array} \\ \left[\begin{array}{ccccc} \frac{\rho_M}{M^2} & \cdots & \frac{-\gamma_M}{M^2} & \cdots & \frac{-\gamma_M}{M^2} \\ \vdots & \vdots & \vdots & \vdots & \vdots \\ 0 & \cdots & \frac{\rho_i}{Mi} & \cdots & \frac{-\gamma_i}{Mi} \\ \vdots & \vdots & \vdots & \vdots & \vdots \\ 0 & \cdots & 0 & \cdots & \frac{-\rho_2}{2M} \end{array}\right] \end{array}$$

Figure 12. Entries in $I - Q^*$ matrix, uniform distribution, constant t.

Other entries can be derived in a similar way. For $i = 2$, using (27) and q_{ij} with $\pi_i = 1/M$ and $p_i = i/M$, and $t_{ij} = t$ when $j \geqslant i$, gives

$$\begin{aligned} u_2 &= \frac{1}{a_{22}} \\ &= \frac{2M}{\rho_2} \end{aligned}$$

and for $i = 3$, the pattern begins to emerge,

$$\begin{aligned} u_3 &= \frac{1 - a_{32}u_2}{a_{33}} \\ &= \frac{3M}{\rho_3} + \frac{\gamma_3}{\rho_3}\left\{\frac{2M}{\rho_2}\right\} \\ &= \frac{3M}{\rho_3} + \frac{M\gamma_3}{\rho_3}\left\{\frac{2}{\rho_2}\right\} \end{aligned}$$

To show the general form of u_i, an induction argument is used. For $l = 2, \ldots, i - 1$ assume the following is true:

$$u_l = \frac{lM}{\rho_l} + \frac{M\gamma_l}{\rho_l}\left\{\frac{l-1}{\rho_{l-1}} + \sum_{r=2}^{l-2} \frac{r}{\rho_r} \prod_{n=r+1}^{l-1}\left(1 + \frac{\gamma_n}{\rho_n}\right)\right\}. \tag{28}$$

We want to prove that for $l = i$ the formula is also true, i.e.

$$u_i = \frac{iM}{\rho_i} + \frac{M\gamma_i}{\rho_i}\left\{\frac{i-1}{\rho_{i-1}} + \sum_{j=2}^{i-2} \frac{j}{\rho_j} \prod_{k=j+1}^{i-1}\left(1 + \frac{\gamma_k}{\rho_k}\right)\right\}. \tag{29}$$

Using (27) gives

$$u_i = \frac{1}{a_{ii}} - \frac{1}{a_{ii}} \sum_{l=2}^{i-1} a_{il}u_l \tag{30}$$

$$= \frac{iM}{\rho_i} + \frac{iM}{\rho_i} \sum_{l=2}^{i-1} \frac{\gamma_i}{iM} u_l \tag{31}$$

$$= \frac{iM}{\rho_i} + \frac{\gamma_i}{\rho_i} \sum_{l=2}^{i-1} u_l. \tag{32}$$

Use Equation (28) to write out the terms for $\sum_{l=2}^{i-1} u_l$ in Equation (32) as follows,

$$
\begin{aligned}
\sum_{l=2}^{i-1} u_l = \frac{2M}{\rho_2} &+ \\
&+\frac{3M}{\rho_3} + \frac{M\gamma_3}{\rho_3}\left\{\frac{2}{\rho_2}\right\} + \\
&+\frac{4M}{\rho_4} + \frac{M\gamma_4}{\rho_4}\left\{\frac{3}{\rho_3} + \frac{2}{\rho_2}\prod_{n=3}^{3}\left(1+\frac{\gamma_n}{\rho_n}\right)\right\} + \\
&+\frac{5M}{\rho_5} + \frac{M\gamma_5}{\rho_5}\left\{\frac{4}{\rho_4} + \frac{2}{\rho_2}\prod_{n=3}^{4}\left(1+\frac{\gamma_n}{\rho_n}\right) + \frac{3}{\rho_3}\prod_{n=4}^{4}\left(1+\frac{\gamma_n}{\rho_n}\right)\right\} + \\
&\vdots \\
&+\frac{(i-2)M}{\rho_{i-2}} + \frac{M\gamma_{i-2}}{\rho_{i-2}}\left\{\frac{i-3}{\rho_{i-3}} + \frac{2}{\rho_2}\prod_{n=3}^{i-3}\left(1+\frac{\gamma_n}{\rho_n}\right) + \frac{3}{\rho_3}\prod_{n=4}^{i-3}\left(1+\frac{\gamma_n}{\rho_n}\right) + \right.\\
&\left.\cdots + \frac{i-4}{\rho_{i-4}}\prod_{n=i-3}^{i-3}\left(1+\frac{\gamma_n}{\rho_n}\right)\right\} + \\
&+\frac{(i-1)M}{\rho_{i-1}} + \frac{M\gamma_{i-1}}{\rho_{i-1}}\left\{\frac{i-2}{\rho_{i-2}} + \frac{2}{\rho_2}\prod_{n=3}^{i-2}\left(1+\frac{\gamma_n}{\rho_n}\right) + \frac{3}{\rho_3}\prod_{n=4}^{i-2}\left(1+\frac{\gamma_n}{\rho_n}\right) + \right.\\
&\left.\cdots + \frac{i-4}{\rho_{i-4}}\prod_{n=i-3}^{i-2}\left(1+\frac{\gamma_n}{\rho_n}\right) + \frac{i-3}{\rho_{i-3}}\prod_{n=i-2}^{i-2}\left(1+\frac{\gamma_n}{\rho_n}\right)\right\}.
\end{aligned}
$$

Collecting terms gives the sum as

$$
\begin{aligned}
&\frac{2M}{\rho_2}\left\{1+\frac{\gamma_3}{\rho_3}+\frac{\gamma_4}{\rho_4}\prod_{n=3}^{3}\left(1+\frac{\gamma_n}{\rho_n}\right)+\frac{\gamma_5}{\rho_5}\prod_{n=3}^{4}\left(1+\frac{\gamma_n}{\rho_n}\right)+\cdots+\frac{\gamma_{i-2}}{\rho_{i-2}}\prod_{n=3}^{i-3}\left(1+\frac{\gamma_n}{\rho_n}\right)+\right.\\
&\left.+\frac{\gamma_{i-1}}{\rho_{i-1}}\prod_{n=3}^{i-2}\left(1+\frac{\gamma_n}{\rho_n}\right)\right\} + \frac{3M}{\rho_3}\left\{1+\frac{\gamma_4}{\rho_4}+\frac{\gamma_5}{\rho_5}\prod_{n=4}^{4}\left(1+\frac{\gamma_n}{\rho_n}\right)+\cdots+\right.\\
&\left.+\frac{\gamma_{i-2}}{\rho_{i-2}}\prod_{n=4}^{i-3}\left(1+\frac{\gamma_n}{\rho_n}\right)+\frac{\gamma_{i-1}}{\rho_{i-1}}\prod_{n=4}^{i-2}\left(1+\frac{\gamma_n}{\rho_n}\right)\right\} + \\
&\vdots \\
&+\frac{(i-3)M}{\rho_{i-3}}\left\{1+\frac{\gamma_{i-2}}{\rho_{i-2}}+\frac{\gamma_{i-1}}{\rho_{i-1}}\prod_{n=i-2}^{i-2}\left(1+\frac{\gamma_n}{\rho_n}\right)\right\}+ \\
&+\frac{(i-2)M}{\rho_{i-2}}\left\{1+\frac{\gamma_{i-1}}{\rho_{i-1}}\right\} + \frac{(i-1)M}{\rho_{i-1}}.
\end{aligned}
$$

Using the following fact,

$$\prod_{n=j}^{i-1}\left(1+\frac{\gamma_n}{\rho_n}\right)=1+\frac{\gamma_j}{\rho_j}+\frac{\gamma_{j+1}}{\rho_{j+1}}\prod_{n=j}^{j}\left(1+\frac{\gamma_n}{\rho_n}\right)+\frac{\gamma_{j+2}}{\rho_{j+2}}\prod_{n=j}^{j+1}\left(1+\frac{\gamma_n}{\rho_n}\right)$$
$$+\cdots+\frac{\gamma_{i-1}}{\rho_{i-1}}\prod_{n=j}^{i-2}\left(1+\frac{\gamma_n}{\rho_n}\right)$$

gives

$$\sum_{l=2}^{i-1}u_l=\frac{2M}{\rho_2}\prod_{n=3}^{i-1}\left(1+\frac{\gamma_n}{\rho_n}\right)+\frac{3M}{\rho_3}\prod_{n=4}^{i-1}\left(1+\frac{\gamma_n}{\rho_n}\right)+\cdots+\frac{(i-3)M}{\rho_{i-3}}\prod_{n=i-2}^{i-1}\left(1+\frac{\gamma_n}{\rho_n}\right)+$$
$$+\frac{(i-2)M}{\rho_{(i-2)}}\prod_{n=i-1}^{i-1}\left(1+\frac{\gamma_n}{\rho_n}\right)+\frac{(i-1)M}{\rho_{i-1}}$$
$$=M\left\{\sum_{j=2}^{i-2}\frac{j}{\rho_j}\prod_{n=j+1}^{i-1}\left(1+\frac{\gamma_k}{\rho_k}\right)+\frac{i-1}{\rho_{i-1}}\right\}.$$

Substituting into Equation (32), gives

$$u_i=\frac{iM}{\rho_i}+\frac{M\gamma_i}{\rho_i}\left\{\frac{i-1}{\rho_{i-1}}+\sum_{j=2}^{i-2}\frac{j}{\rho_j}\prod_{n=j+1}^{i-1}\left(1+\frac{\gamma_k}{\rho_k}\right)\right\}$$

as required (see Equation (29)).

For the special case when $i=M$ the expected number of iterations to absorption, starting in the worst state y_M, in terms of p and M is,

$$u_M=\frac{M^2}{\rho_M}+\frac{M\gamma_M}{\rho_M}\left\{\frac{M-1}{\rho_{M-1}}+\sum_{j=2}^{M-2}\frac{j}{\rho_j}\prod_{k=j+1}^{M-1}\left(1+\frac{\gamma_k}{\rho_k}\right)\right\}.$$

Assuming $\prod_{k=i+1}^{M-1}(1+\frac{\gamma_k}{\rho_k})=1$ if $i\geqslant M-1$, u_M can be written as

$$u_M=\frac{M^2}{\rho_M}+\frac{M\gamma_M}{\rho_M}\left\{\sum_{j=2}^{M-1}\frac{j}{\rho_j}\prod_{k=j+1}^{M-1}\left(1+\frac{\gamma_k}{\rho_k}\right)\right\}$$

which can also be written as

$$u_M=1+\sum_{j=2}^{M}\frac{j}{\rho_j}\prod_{k=j+1}^{M}\left(1+\frac{\gamma_k}{\rho_k}\right).$$

For the special case when $i = 2$ we get directly from (13) that

$$u_2 = \frac{2M}{\rho_2}. \qquad \square$$

THEOREM 6. *The probability of absorption in state y_1 for the embedded Markov chain model of PAS/PRS, assuming a uniform distribution for the generating method and starting in state y_M is*

$$f_{M1} = \frac{\gamma_M}{\rho_M}\left\{1 + \sum_{j=2}^{M-1} \frac{\gamma_j}{\rho_j} \prod_{k=j+1}^{M-1}\left(1 + \frac{\gamma_k}{\rho_k}\right)\right\}, \tag{33}$$

where p is the probability of sampling according to PAS, t is an acceptance probability and γ_k and ρ_k are defined as

$$\gamma_k = pM + k(1-p),$$

$$\rho_k = (k-1)\left(pM + k(1-p) + \left(\frac{k}{k-1}\right)(1-p)(M-k)t\right).$$

For the special case when starting in state y_2 the probability of absorption in state y_1 is

$$f_{21} = \frac{\gamma_2}{\rho_2}. \tag{34}$$

PROOF. Theorem 2, Equation (7) gives F as $F = (I - Q^*)^{-1}R$, and f_{M1} is the $M1$th element in the F matrix. The $(I-Q^*)^{-1}$ matrix for a uniform distribution is shown in Figure 13. It is easy to verify that this is the inverse of $(I - Q^*)$ shown in Figure 12, by checking that $(I - Q^*)(I - Q^*)^{-1} = I$, where I is the identity matrix. The first column of R is given by $[\frac{\gamma_M}{M^2}, \ldots, \frac{\gamma_i}{iM}, \ldots, \frac{\gamma_2}{2M}]^{\mathrm{T}}$. Multiplying the Mth row of the $(I - Q^*)^{-1}$ matrix with the first column of R gives f_{M1} as follows,

$$f_{M1} = \frac{\gamma_M}{\rho_M} + \frac{\gamma_M}{\rho_M}\left\{\sum_{j=2}^{M-2} \frac{\gamma_j}{\rho_j} \prod_{k=j+1}^{M-1}\left(1 + \frac{\gamma_k}{\rho_k}\right) + \frac{\gamma_{M-1}}{\rho_{M-1}}\right\}.$$

Assuming $\prod_{k=i+1}^{M-1}(1 + \frac{\gamma_k}{\rho_k}) = 1$ if $i \geqslant M-1$, f_{M1} can be written as

$$f_{M1} = \frac{\gamma_M}{\rho_M}\left\{1 + \sum_{j=2}^{M-1} \frac{\gamma_j}{\rho_j} \prod_{k=j+1}^{M-1}\left(1 + \frac{\gamma_k}{\rho_k}\right)\right\}.$$

To determine f_{21} we multiply the row corresponding to state 2 of the $(I - Q^*)^{-1}$ matrix with the first column of R, giving $f_{21} = \gamma_2/\rho_2$. $\square$

$$
\begin{array}{c}
\\ M \\ M-1 \\ \vdots \\ \\ i \\ \\ \vdots \\ 2
\end{array}
\begin{array}{c}
\begin{array}{cccccc} M & M-1 & \cdots & i & \cdots & 2 \end{array} \\
\left[\begin{array}{cccccc}
\frac{M^2}{\rho_M} & \frac{M\gamma_M(M-1)}{\rho_M\rho_{M-1}} & \cdots & \frac{M\gamma_M i}{\rho_M\rho_i}\prod_{k=i+1}^{M-1}(1+\frac{\gamma_k}{\rho_k}) & \cdots & \frac{M\gamma_M 2}{\rho_M\rho_2}\prod_{k=3}^{M-1}(1+\frac{\gamma_k}{\rho_k}) \\
0 & \frac{M(M-1)}{\rho_{M-1}} & \cdots & \frac{M\gamma_{M-1} i}{\rho_{M-1}\rho_i}\prod_{k=i+1}^{M-2}(1+\frac{\gamma_k}{\rho_k}) & \cdots & \frac{M\gamma_{M-1} 2}{\rho_{M-1}\rho_2}\prod_{k=3}^{M-2}(1+\frac{\gamma_k}{\rho_k}) \\
\vdots & \vdots & \vdots & \vdots & \vdots & \vdots \\
0 & 0 & \cdots & \frac{iM}{\rho_i} & \cdots & \frac{M\gamma_i 2}{\rho_i\rho_2}\prod_{k=3}^{i-1}(1+\frac{\gamma_k}{\rho_k}) \\
\vdots & \vdots & \vdots & \vdots & \vdots & \vdots \\
0 & 0 & \cdots & 0 & \cdots & \frac{2M}{\rho_2}
\end{array}\right]
\end{array}
$$

Figure 13. Entries in $(I - Q^*)^{-1}$, uniform distribution, constant t.

Bibliography

[1] Kirkpatrick, S., Gelatt Jr., C. D. and Vecchi, M.P.: Optimization by simulated annealing, *Science* **20** (1983), 671–680.

[2] Metropolis, N., Rosenbluth, A., Rosenbluth, M., Teller, A. and Teller, E.: Equation of state calculations by fast computing machines. *J. Chem. Phys.* **21** (1953), 1087–1090.

[3] Zabinsky, Z. B. and Kristinsdottir, B. P.: *Complexity Analysis Integrating Pure Adaptive Search (PAS) and Pure Random Search (PRS)*, Developments in Global Optimization, edited by I. M. Bomze et al., 1997, 171–181.

[4] Romeijn, E. H. and Smith, R. L.: Simulated annealing and adaptive search in global optimization, *Probab. Eng. Inform. Sci.* **8** (1994), 571–590.

[5] Bulger, D. W. and Wood, G. R., Hesitant adaptive search for global optimisation, *Math. Programming* **81** (1998), 89–102.

[6] Wood, G. R., Zabinsky, Z. B. and Kristinsdottir, B. P.: Hesitant adaptive search: The distribution of the number of iterations to convergence, *Math. Programming* **89**(3) (2001), 479–486.

[7] Aarts, E. and Korst, J.: *Simulated Annealing and Boltzmann Machines: A Stochastic Approach to Combinatorial Optimization and Neural Computing*, Wiley, New York, 1989.

[8] Rudolph, G.: Convergence analysis of canonical genetic algorithms, *IEEE Trans. Neural Networks* **5** (1994), 96–101.

[9] Kemeny, J. G. and Snell, J. L.: *Finite Markov Chains*, Springer-Verlag, New York, 1976.

[10] Patel, N. R., Smith, R. L. and Zabinsky, Z. B.: Pure adaptive search in Monte Carlo optimization, *Math. Programming* **43** (1988), 317–328.

[11] Zabinsky, Z. B. and Smith, R. L.: Pure adaptive search in global optimization, *Math. Programming* **53** (1992), 323–338.

[12] Zabinsky, Z. B., Wood, G. R., Steel, M. A. and Baritompa, W. P.: Pure adaptive search for finite global optimization, *Math. Programming* **69** (1995), 443–448.

[13] Ravindran, A., Phillips, D. T. and Solberg, J. J.: *Operations Research, Principles and Practice*, Wiley, 1976.

[14] Kristinsdottir, B. P: Complexity analysis of random search algorithms, Ph.D. Dissertation, University of Washington, 1997.

Chapter 10

PARALLEL BRANCH-AND-BOUND ATTRACTION BASED METHODS FOR GLOBAL OPTIMIZATION

Kaj Madsen
Department of Mathematical Modeling
Technical University of Denmark
Lyngby, Denmark
km@imm.dtu.dk

Julius Žilinskas
Department of Informatics
Kaunas University of Technology
Studentu 50-214b, Kaunas, Lithuania
jzil@dsplab.ktu.lt

Abstract In this paper a parallel version of an attraction based branch-and-bound method for global optimization is presented. The method has been implemented and tested using a parallel Scali system. Some well known test functions as well as two practical problems were used for the testing. The results show the prospectiveness of dynamic load balancing for the distributed parallelization of the considered algorithm.

Keywords: Global optimization, parallel branch-and-bound, testing of GO algorithms

1. Introduction

The attraction based multidimensional branch-and-bound method for global optimization was proposed in [1]. The method extends successful algorithmic ideas of the interval arithmetic based methods to the case where calculation of an objective function takes place in a black box, i.e. the algorithm cannot utilize information on how function values are found. Since rigorous calculation of inclusion intervals for the values of an objective function is impossible in the black box situation, prospectiveness of different subregions with respect

G. Dzemyda et al. (eds.), Stochastic and Global Optimization, 175–187.
© 2002 *Kluwer Academic Publishers.*

to finding a global minimizer is evaluated using the information gathered during local descents. The stationary points are evaluated by means of local descents starting from regular and/or random initial points. An approximate lower bound of the objective function in a subregion is evaluated using the values of the objective function and its gradient (or finite difference approximations of the gradient) calculated during the local searches. A subregion is either excluded from the further search or chosen for further subdivision depending on information on stationary points and on the lower bound for the objective function values. The method, briefly presented in Section 2, is heuristic in the sense that, unlike the interval methods, it does not guarantee to find the global minimum with a prescribed accuracy.

The performance of the sequential version of the method has been evaluated by means of experimental testing [2]. The results have shown that the method performs quite well not only for the test functions but also for some practical problems. Since global optimization methods are computationally intensive, parallelization is recognized as one of the most prospective computational approaches. Two versions of parallelizing the algorithm are discussed in Section 3.

The Scali parallel system of the Department of Computer Science, University of Copenhagen, was used for experiments with parallel algorithms. The codes were implemented in C++ using the Message Passing Interface which is a portable standardized communication protocol for massively parallel machines. The parallel system is briefly described in Section 4. In Section 5 the testing methodology and the criteria of efficiency for parallel algorithms are discussed. Section 6 describes the experimental results, and in Section 7 some concluding remarks are given and possible directions of future work are discussed.

2. Method

A new branch-and-bound type method for global optimization over a compact right parallelepiped D in $\Re^n$, parallel to the coordinate axes, has been proposed in [1]. The method is based on partitioning of the feasible region controlled by the information about the stationary points and values of the objective function f and its gradients f' (or finite difference approximations) calculated during local searches. It is generally applicable in a black box situation, i.e. function values may be calculated by a black box subroutine.

For some classes of problems, e.g., analytically defined functions with modest number of variables, interval methods have been very successful. The new technique is based on the same general branch-and-bound principle, but without having the need of being rigorous. Thus the guaranteed convergence property of the interval method is lost. The method can be interpreted as a strategy for managing local searches (as, e.g., quasi-Newton methods) in a branch-and-bound search for global minimizers. Once a local minimizer z has been found a domain around z is left out from further search, thus we often avoid several searches to the same local minimizer.

Having the complexity of the global optimization problem in mind, it is not intended to guarantee that the solution is found, but rather to try to get the best possible result that can be obtained by using a limited amount of computational resources. Therefore, the subregions evaluated as not prospective are not completely excluded from the search but stored in the "garbage store" for further analysis if the computing resources will allow such an analysis. The user should specify how much storage and computer time can be afforded rather than specifying some accuracy tolerance as required by many traditional approaches. If the time allows the search is restarted using the information on subregions in the "garbage store".

Since the solution may be lost the model algorithm is embedded into a loop, which will be referenced as the outer loop. Until the predefined computing time will expire, there exists the possibility to restart the search in the boxes kept as the *outer* candidate set G, in addition to the *inner* candidate set C processed by the model algorithm. When the latter has been finished the elements of G will cover D except for the balls around the known stationary points.

In the outer iteration G forms the basis for a re-generation of the inner candidate set C: For each box B in G we use the available information about the stationary points of B in a device which splits B into several boxes that are all added to C. Furthermore parameters which determine the number of points used during the tests of the reduction phase in the inner iteration may be changed from the outer iteration in order to increase the reliability of the method.

The structure of the algorithm is shown in Figure 1. Local searches, denoted *Newton(B)* in Figure 2 presenting the sub-algorithm *reduce-or-subdivide*, make essential part of the computational work. Normally local searches only guarantee convergence to *stationary points* and therefore we will use this phrase instead of *local minimizer*. When a stationary point z has been found in B, a ball

$$ball_z = \{x \in B \mid \|x - z\| \leqslant \varepsilon_{\text{cluster}}\} \tag{1}$$

is left out from further consideration, where $\varepsilon_{\text{cluster}}$ is a user provided parameter. Thus the candidate set consists of boxes where balls with known stationary points have been left out. For ease of notation we still denote such a subdomain a *box*. Besides of local searches *reduce-or-subdivide* performes calculation of the lower bound for function values. The estimate of global minimum, denoted $\bar{f}$ is equal to the best local minimum found.

The boolean *inner-stop* is true when C contains no boxes (i.e. only points). The outer stopping condition is a time limit set by the user. Other stopping

```
initialize: C, f̄, G
while not time-limit exceeded do
        while not inner-stop do
                remove-best(C) → B
                generate sample points
                reduce-or-subdivide(B) → result, f̄, garbage
                C ∪ {result} → C
                G ∪ {garbage} → G
        end
        {p ∈ C | f(p) ⩽ f̄ + μ} → SolSet
        while G ≠ ∅ do
                remove-one(G) → B
                subdivide(B) → result
                C ∪ {result} → C
        end
        perhaps adjust reduction parameters
end
```

Figure 1. Structure of the algorithm.

```
if monotone then
        Monotone(B)→ result
        B → garbage
else    if (case 1) or (case 2) then
                Newton(B)→ result
                B \ {known stationary points} → garbage
        else    if (case 3) and (LB(B) > f̄) then
                        ∅ → result
                        B \ {known stationary points} → garbage
                else    subdivide(B)→ B¹, B²
                        {B¹, B²} \ {known stationary points} → result
                        ∅ → garbage
update f̄
```

Figure 2. Reduce-or-subdivide. The phrase {known stationary points} means $\{ball_z \mid z \in B$ is a known stationary point$\}$

conditions might be adopted in case of additional information on the problem. The variable *SolSet* is intended to converge towards the set of global minimizers, X^*. μ is a positive number which is intended to distinguish X^* from other stationary points. In theory μ might be 0, but in practice (i.e. when rounding errors are present) μ has to be positive in order that *SolSet* eventually will contain the whole of X^*.

The details of the algorithm are now explained.

- *remove-best*(C): The box B from C which has the smallest known function value is chosen.

- *Sample points.* The sample points are used in the monotonicity and Newton tests, see below. Two strategies are applied for choosing these points. First we use the following regular distribution of up to $2n+1$ points: One point at $m(B)$, the center of B, and two points for each coordinate direction, $m(B) \pm \frac{1}{3} w(B^{(j)}) * e_j$, $j = 1, \ldots, n$, where $B = B^{(1)} \times \cdots \times B^{(n)}$, $w(B)$ means width of the box B, and e_j is the j-th unit vector. The constant $1/3$ is chosen because it provides the best coverage of B in the sense of disjoint balls with centers at the regular sample points.
  Secondly we may use random points, uniformly distributed in B.
  The total number of sample points, $p_1 = n_{\text{sample}}$, has to be provided by the user. The sample points are chosen in the order they are mentioned here, i.e. we only use random points if n_{sample} is greater than $2n + 1$.

- *reduce-or-subdivide.* The reduce-or-subdivide procedure is summarized in Figure 2, and the explanation follows:

  - *Monotonicity test.* This test is based on gradient information at the sample points (explicitly calculated if possible, by means of finite differences otherwise). If a stationary point has previously been found in B, however, then f cannot be strictly monotonous in B, and hence the test is skipped.
    Otherwise we examine the gradients. If there exists at least one coordinate x_i for which $\partial f(x)/\partial x_i$ has constant sign at all sample points then we decide that no stationary point exists in B. The result of the test *Monotone*(B) is equal to $B \cap \partial_i(D)$ where $\partial_i(D)$ denotes the facet of D, which is orthogonal to x_i axis, and towards which f decreases. The variable *garbage* is set to B.

  - *Newton test.* This is performed if the *Monotonicity test* has not reduced B. Local searches for finding stationary points of f in B are started from each sample point. Gradients are estimated using forward differences. Notice that balls, $ball_z$, around stationary points z which are known at the beginning of the current Newton test are considered "outside of B".
    We distinguish the following four cases:
    Case 1. Every iteration sequence has an iterate which is outside of B. In this case we decide that no stationary point exists in B, and the result of the test *Monotone*(B) is equal to $B \cap \partial(D)$ where $\partial(D)$ denotes the border of D.

Case 2. All iteration sequences provide convergence to the same point $z \in B$. Then the result of the test $Monotone(B)$ is equal to $\{z\}$. (In practice each Newton iteration is stopped by using one of the usual stopping rules for local searches (to prevent infinite iterations), and two stopping points are considered equal if their distance is less than $\varepsilon_{\text{cluster}}$.)
Case 3. Several stationary points have been found in B. In this case the lower bound reduction may take place. The lower bound $LB(B)$ is estimated using the maximal gradient norm *maxgrad* obtained during the local searches in B. Let $f(x_{\min})$ be the smallest known function value in B, then

$$LB(B) = \min\{f(x_{\min}) - maxgrad * \|x - x_{\min}\| \mid x \in B\}.$$

If $LB(B) > \bar{f}$ then $\emptyset \rightarrow$ *result* otherwise subdivision takes place. (We decided only to use the lower bound reduction in Case 3 because our experiments indicate that this is more efficient.)
Case 4. None of the first 3 cases has occurred and subdivision takes place.

- *Subdivision*. We always split B into two by a hyperplane which is orthogonal to one of the coordinate directions. *result* consists of the two new parts, and $\emptyset \rightarrow$ *garbage*. The subdivision is done as follows:
  Case a. No known stationary point in B: The splitting divides B into two equal parts separated by a hyperplane through the center of B and perpendicular to the side of maximal length.
  Case b. One known stationary point in B: Find the coordinate for which the distance from this point to a side of B is maximal. The splitting plane is chosen to be perpendicular to this coordinate axis, and it halves the distance mentioned.
  Case c. More than one known stationary points in B: Find the two stationary points with smallest function values. Find the coordinate for which the distance between these two points is maximal. The splitting is made with the plane which is perpendicular to this coordinate direction and has the same distance to the two stationary points.

- G is emptied in the second inner **while**-loop of algorithm. After B has been removed from G by the *remove-one* procedure it is subdivided into two parts by the *Subdivision* procedure.

Notice that if the local search has the following property: "If there is no stationary point in B then eventually an iterate outside of B will be generated", then boxes without stationary points will always lead to Case 1. Thus the number of boxes in C with full dimensionality and without known stationary points must be very limited. Hence the monotonicity test is most often skipped. Furthermore, the distance between known stationary points is at least $\varepsilon_{\text{cluster}}$.

3. Parallel Algorithms

A parallel version of a branch-and-bound algorithm may correspond to one of two main paradigms of parallel programming: master-slave and distributed. In the parallel master-slave system there is one master processor, which controls the optimization process. The slave processors receive subproblems from the master, perform computations on them and send the results to the master. In this case it is possible to ensure that all slaves perform computations on promising subproblems, the master controls load balancing and termination. The disadvantage of this scheme is that master can become a bottleneck especially when the system is large and the communications are frequent and heavy. The slaves can become idle when waiting for subproblems in this case.

The master-slave implementation of the method is straight forward. The master processor runs the algorithm similarly to the sequential algorithm from Figure 1: the subproblems are sent to slave processors, those perform local searches, subdivide subproblems and send the results back to the master. The package of the data on a subproblem should include information on its stationary points found in the corresponding box. Thus the communication is intensive and involves much data. The interesting issue is the trade-off between the quantity of communicated information and calculation time. The master-slave implementations of the method are described and the results of the experiments are discussed in [3]. The master-slave paradigm is suitable when the number of processors is not large.

The second paradigm, i.e. distributed parallelization, assumes a cooperative work of a set of equal communicating processors. The main problems of the distributed parallelization are load balancing and termination detection.

There are two different balancing strategies: static and dynamic. When the static load balancing is used, the subproblems are initially distributed and then the processors work independently and do not exchange any later generated subproblems. The application of static load balancing to the considered problem corresponds to the method of geometric parallelization. Its implementation is simple, but its disadvantage is unpredictable distribution of work load implying situations when some processors become idle.

Dynamic load balancing aims to share the work load uniformly. However, to implement such a strategy the problems of controling the load balance and termination detection should be solved.

The existence of the outer loop involves new aspects in the distributed parallel implementation of the method. Even with static load balancing all processors can always have work to do, but the depth of calculation may differ. Using the dynamic load balancing, however, the depth of calculation (in between processors) may become similar. One way is to perform load balancing having the depth of calculation in mind. Another way is to balance loading so that all processors finish inner loops together and then redistribute subproblems.

Experiments with the serial version of the algorithm have shown [2] that in almost all cases the solution is found during the first run of the outer loop. The investigation of the parallel distributed version of the algorithms has been started with static load balance and without the outer loop. Experiments with such algorithms should show how evenly the amount of work is distributed.

Two versions of such algorithms have been implemented: without and with the exchange in between processors of the currently best known value of the objective function $\bar{f}$. Since the same algorithm should be used in all cases, the same initial subdivision must take place independently of the number of processors. Therefore, the domain D of the problem is always divided into 16 sub-domains, which are distributed to the processors.

4. Parallel System

The Scali system at the Department of Computer Science, University of Copenhagen (DIKU) was used for experiments with the parallel algorithms. A Scali system consists of a control node, a cluster and Scali software. A cluster is a set of processing nodes (computers) interconnected using high speed Scalable Coherent Interface (SCI) interconnect. The Scali software makes the cluster act as a single system – the parallel computer. A front-end control node compiles and launches applications at the processing nodes and acts as a license server of the Scali software. The hardware of DIKU's Scali system consists of a front-end node Kvaser and 16 processing nodes (wulf-node-11 ... wulf-node-44), that are connected into a 4×4 two-dimensional ring (torus) structure using a 3.6 Gbit SCI network. All of the 17 machines are Pentiums running Redhat Linux and they are connected with a 100Mbit fast ethernet. The DIKU Scali system is shown in Figure 3.

All codes were implemented in C++. The Message Passing Interface (MPI) was used for communication in between the parallel processors. MPI is a standardized portable communication protocol used on massively parallel machines. The Scali MPI implementation (ScaMPI) is compliant with the MPI 1.1 specification [4].

5. Criteria of Efficiency of Parallel Algorithms

A commonly used criterion for measuring the performance of parallel algorithms is speedup:

$$s_m = \frac{t_1}{t_m},$$

where t_m is the time used by the algorithm implemented on m processors. The speedup divided by the number of processors is usually called the efficiency:

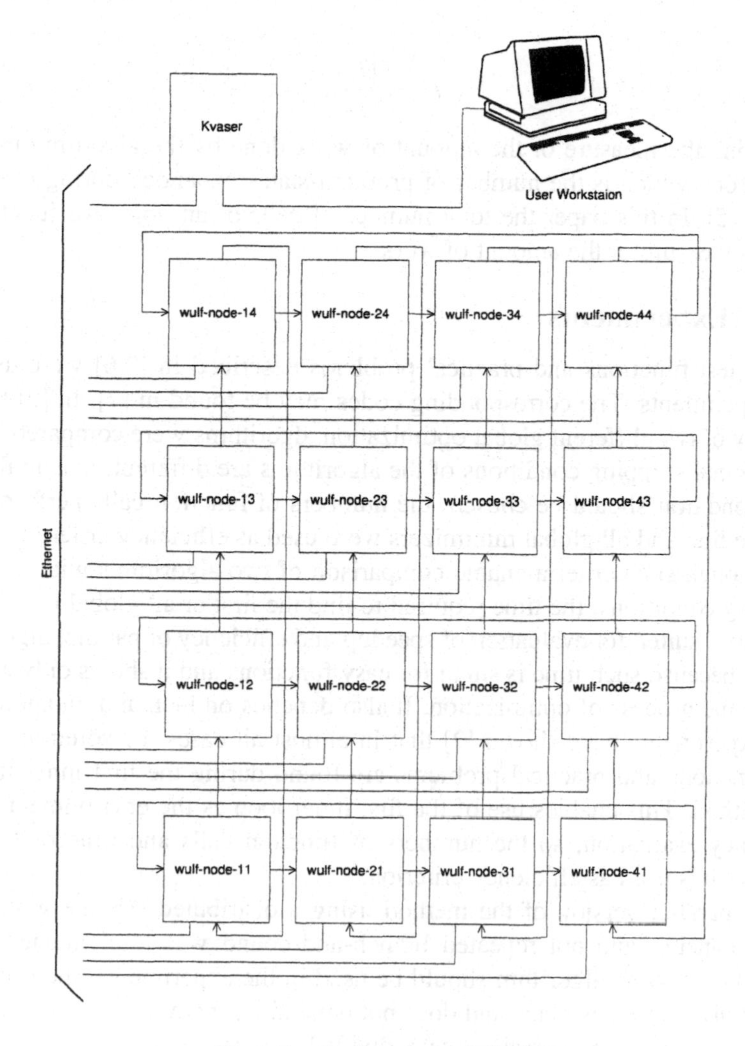

Figure 3. Scali system at DIKU.

$$e_m = \frac{s_m}{m}.$$

Normally $1 \leqslant s_m \leqslant m$ and $0 \leqslant e_m \leqslant 1$, but detrimental ($s_m < 1$), deceleration ($s_{m_1} < s_{m_2}$, for $m_1 > m_2$) and acceleration ($s_m > m$) anomalies [5] can occur in parallel branch-and-bound algorithms, since the progress of the search, and therefore the amount of work, can vary. The pseudo efficiency criterion is suggested in [5]:

$$pe_m = \frac{r_m}{m \times r_1},$$

where

$$r_k = \frac{|T_k|}{t_k},$$

$|T_k|$ being the measure of the amount of work done by the algorithm using k processors, which is the number of problem-states expanded during the solution in [5]. In this paper the total number of calls of an objective function is used as measure of the amount of work.

6. Experiments

The test functions and practical problems described in [2,6] were used in our experiments. The corresponding codes may be found in [7]. In [2] the efficiency of two different global optimization algorithms were compared. Since the original stopping conditions of the algorithms are different, a neutral stopping condition should be chosen; the numbers of function calls performed to find the first and all global minimizers were used as efficiency criterion.

Although such criteria enable comparison of two algorithms with different stopping conditions, the time required to find the first or all global minimizers is not well suited for evaluation of speedup and efficiency of parallel algorithm. This is because such time is small for easy functions and it shows only a small part of the process of optimization. It also depends on luck and on anomalies. The experiments have shown [2] that in almost all cases the solutions of the test functions and practical problems are found during the first inner loop of the method. This enables use of the first inner loop as the environment of the efficiency evaluation, so the numbers of function calls and time of the first inner loop is used as efficiency criterion.

The parallel version of the method using a distributed scheme with static load balancing and not repeated branch-and-bound was implemented using MPI. As the same algorithm should be used in the experiments, the same initial subdivision takes place and does not depend on the number of processors. The domain of a problem is always divided into 16 sub-domains, which are distributed to parallel processors (or processed by one in the sequential case).

The computing times in seconds (t, s) and the numbers of objective (Nrf) and gradient (Nrg) function calls required for the first inner loop were measured during the parallel computations. These values were used for evaluation of the speedup (s_m), efficiency (e_m) and pseudo efficiency (pe_m). The performance of the algorithm when the number of processors is $m = 1$ is given in Table 1. The performance and criteria of paralellization of the algorithm when the numbers of processors are $m = 4$ and $m = 16$ are given in Tables 2 and 3. The results may have been disturbed by other simultaneous users of the parallel system, though.

The numbers of the Levy and Growth functions of [6] are not given here. This is because the first inner loop for these functions lasts too long and the

Table 1. The performance of the sequential algorithm (number of processors is $m = 1$).

Function	n	Nrf	Nrg	t, s
Rosenbrock	2	115	87	0.0018
McCormic	2	85	79	0.0018
Box and Betts	3	425	423	0.043
Paviani	10	447	359	0.021
Gen Ros	30	40487	39185	10.6
Gold and Price	2	212	168	0.0045
Shekel 5	4	1043	771	0.044
Shekel 7	4	931	644	0.049
Shekel 10	4	1371	999	0.099
Griewank	10	4713	3859	0.27
6-HumpC	2	119	103	0.0026
Branin	2	651	516	0.013
Shubert	2	10629	7102	0.26
Hansen	2	9512	6363	0.24
Cola	17	131909	106067	120.5

Table 2. The performance of the algorithm with static load balance when the number of processors is $m = 4$.

Function	n	Nrf	Nrg	t, s	s_4	e_4	pe_4
Rosenbrock	2	366	319	0.0031	0.57	0.14	0.45
McCormic	2	185	180	0.0019	0.97	0.24	0.53
Box and Betts	3	402	400	0.019	2.3	0.57	0.54
Paviani	10	597	595	0.017	1.2	0.31	0.41
Gen Ros	30	8967	8448	1.2	9.0	2.3	0.50
Gold and Price	2	251	208	0.0023	1.9	0.48	0.56
Shekel 5	4	1265	1185	0.023	1.9	0.47	0.57
Shekel 7	4	1314	1146	0.037	1.3	0.33	0.47
Shekel 10	4	719	619	0.029	3.5	0.87	0.45
Griewank	10	7266	6036	0.39	0.68	0.17	0.26
6-HumpC	2	524	448	0.0051	0.51	0.13	0.56
Branin	2	1015	865	0.0086	1.5	0.38	0.60
Shubert	2	10932	7223	0.075	3.4	0.86	0.88
Hansen	2	9352	6201	0.062	3.8	0.95	0.93
Cola	17	70880	58076	26.9	4.5	1.1	0.60

huge number of boxes creates memory problems at the Scali system. These difficulties suggest to improve the method in such way that the boxes which are unlikely to contain the global minimizer would be discarded and not kept in the outer candidate set. Some improvements may be achieved by introducing of the more tight lower bounds.

The progress of the search may differ for the different numbers of processors. Therefore the amount of work needed to finish the first inner loop may also be different. If a good value of the objective function is found earlier then more boxes are eliminated and put into garbage list. On the other hand if the good value is found later then more boxes are explored and the amount of work

Table 3. The performance of the algorithm with static load balance when the number of processors is $m = 16$.

Function	n	Nrf	Nrg	t, s	s_{16}	e_{16}	pe_{16}
Rosenbrock	2	367	356	0.0024	0.76	0.047	0.15
McCormic	2	417	414	0.0020	0.90	0.056	0.28
Box and Betts	3	539	535	0.010	4.1	0.25	0.32
Paviani	10	1479	1478	0.0071	2.9	0.18	0.61
Gen Ros	30	15332	11538	1.5	6.9	0.43	0.16
Gold and Price	2	528	498	0.0024	1.9	0.12	0.29
Shekel 5	4	1652	1576	0.036	1.2	0.076	0.12
Shekel 7	4	1406	1348	0.019	2.5	0.16	0.24
Shekel 10	4	1345	1272	0.025	4.0	0.25	0.25
Griewank	10	9562	8444	0.39	0.67	0.042	0.085
6-HumpC	2	736	668	0.021	0.12	0.0076	0.047
Branin	2	1613	1449	0.34	0.039	0.0024	0.0060
Shubert	2	11445	7878	0.16	1.6	0.099	0.11
Hansen	2	12474	8730	0.44	0.54	0.034	0.045
Cola	17	12994	11638	7.8	15.5	0.97	0.096

is larger. That are called detrimental, deceleration or acceleration anomalies in [5]. In our experiments the detrimental anomaly takes place when minimizing the Rosenbrock, McCormic, Griewank and 6-Hump Camel functions for $m = 4$ and $m = 16$, and the Branin and Hansen functions for $m = 16$. Acceleration anomalies takes place when minimizing the Generalized Rosenbrock function and the Cola problem, for $m = 4$. Even when speedup and efficiency are in the proper ranges: $1 \leqslant s_m \leqslant m$ and $0 \leqslant e_m \leqslant 1$, these problems do not show real improvement. For example, the speedup is close to the number of processors and the efficiency is 97% when the Cola problem is minimized using 16 processors, but many of the processors are idle for most of the time. Therefore speedup and efficiency are not good criteria in our case. The pseudo efficiency depends on the amount of work and is more suited for the evaluation of parallelization.

As it is seen from the tables, the value of the pseudo efficiency is best (largest) when the number of processors is $m = 4$. This may be because the amount of work is distributed more evenly when each processor has more than just one box initially. The experimental results indicate that a different load balancing strategy must be considered.

However the results for $m = 4$ show that parallelization is useful (the pseudo efficiency is more than 50% in almost all cases) and good results will probably be achieved when the load balancing is improved.

7. Conclusions and Future Work

The experimental results of the parallel implementation of the method have proved the usefulness of parallelization in this case: Computing times are decreased in most cases. However the pseudo efficiency is rather bad for 16 parallel processors, and improvements also seem possible for 4 parallel processors. One way to obtain better results will be by improving the load balance. Hence a primary direction of the future research is implementation of a dynamic load balance and further experiments with parallel implementations of the method.

Bibliography

[1] Madsen, K.: Real versus interval methods for global optimization, Presentation at the Conference *Celebrating the 60th Birthday of M. J. D. Powell*, Cambridge, July 1996.

[2] Madsen, K. and Žilinskas, J.: Evaluating performance of attraction based subdivision method for global optimization, In: *Second International Conference "Simulation, Gaming, Training and Business Process Reengineering in Operations"*, RTU, Latvia, 2000, pp. 38–42.

[3] Gudmundsson, S.: *Parallel Global Optimization*, M.Sc. Thesis, IMM, Technical University of Denmark, 1998.

[4] Message Passing Interface Forum. MPI: A Message-Passing Interface standard (version 1.1), Technical Report, 1995. http://www.mpi-forum.org.

[5] Rayward-Smith, V. J., Rush, S. A. and McKeown, G. P.: Efficiency considerations in the implementation of parallel branch-and-bound, *Ann. Oper. Res.* **43** (1993), 123–145.

[6] Madsen, K. and Žilinskas, J.: Testing of branch-and-bound methods for global optimization, IMM-REP-2000-05, Department of Mathematical Modelling, Technical University of Denmark, DK-2800 Lyngby, Denmark, 2000.

[7] Madsen, K.: Test problems for global optimization, http://www.imm.dtu.dk/~km/GlobOpt/testex/

Chapter 11

ON SOLUTION OF STOCHASTIC LINEAR PROGRAMS BY DISCRETIZATION METHODS

K. Marti
Federal Armed Forces University
Aero-Space Engineering and Technology
D-85577 Neubiberg, München, Germany
kurt.marti@unibw-muenchen.de

Abstract Stochastic linear programs (SLP) with complete fixed recourse are solved approximatively by means of discretization of the underlying probability distribution of the random parameters. Error estimates are given, and a priori bounds for the approximation error are derived. Furthermore, exploiting invariance properties of the probability distribution of the random parameters, problem-oriented discretizations are derived which simplify then the computation of admissible descent directions at non-stationary points.

Keywords: Stochastic linear programs (SLP), discretization methods, a priori error bounds, invariant probability distributions

1. A priori Error Bounds

A well-known method to handle linear programs

$$\text{minimize } c(\omega)'x \quad \text{s.t. } T(\omega)x = h(\omega), \quad x \in D, \tag{1}$$

with random data $(c(\omega), T(\omega), h(\omega))$ on a probability space $(\Omega, \mathcal{A}, P)$ is to replace (1), cf. [1], by the stochastic optimization problem

$$\text{minimize } F(x) \quad \text{s.t. } x \in D, \tag{2a}$$

where

$$F(x) = E\big(c(\omega)'x + p\big(h(\omega) - T(\omega)x\big)\big) \tag{2b}$$

and $p = p(z)$ denote the so-called second stage costs defined by

$$p(z) = \inf\{q'y : Wy = z, y \geqslant 0\}, \quad z \in \mathbb{R}^m. \tag{2c}$$

G. Dzemyda et al. (eds.), Stochastic and Global Optimization, 189–207.
© 2002 *Kluwer Academic Publishers.*

Here, D is a fixed convex polyhedron in $\mathbb{R}^n$, hence, "$x \in D$" represents the deterministic constraints in (1), and "E" denotes the expectation operator. In the following we suppose that the $m \times n_1$ matrix W and the m vector q are related such that

$$\{Wy : y \geqslant 0\} = \mathbb{R}^m, \tag{3a}$$

$$\{v : W'v \leqslant q\} \neq \emptyset. \tag{3b}$$

Thus, the loss function p is defined on the whole $\mathbb{R}^m$. If $q \geqslant 0$, then (3b) holds, and p is nonnegative. According to [2] we know that p is a sublinear function on $\mathbb{R}^m$, thus

$$p(z+w) \leqslant p(z) + p(w) \quad \text{for all } z, w \in \mathbb{R}^m, \tag{4a}$$

$$p(\lambda z) = \lambda p(z) \quad \text{for all } z \in \mathbb{R}^m, \lambda \geqslant 0. \tag{4b}$$

Consequently, p is a convex function on $\mathbb{R}^m$, and we have that

$$p(0) = 0, \tag{4c}$$

$$-p(-z) \leqslant p(z) \quad \text{for all } z \in \mathbb{R}^m, \tag{4d}$$

$$-p(z-w) \leqslant p(w) - p(z) \leqslant p(w-z) \quad \text{for all } z, w \in \mathbb{R}^m. \tag{4e}$$

Moreover, if

$$\|p\| = \sup_{\|z\|_E \leqslant 1} |p(z)| = \sup_{\|z\|_E = 1} |p(z)| \quad (< +\infty) \tag{4f}$$

denotes the norm of the sublinear function p, then

$$|p(z)| \leqslant \|p\| \|z\| \quad \text{for all } z \in \mathbb{R}^m, \tag{4g}$$

$$|p(w) - p(z)| \leqslant \|p\| \|w - z\| \quad \text{for all } z, w \in \mathbb{R}^m. \tag{4h}$$

Denoting $Ec(\omega)$ by $\bar{c}$, (2b) reads

$$F(x) = \bar{c}'x + Ep\big(h(\omega) - T(\omega)x\big). \tag{5}$$

2. Disretization and Error Bounds

Approximating now the $m \times (n+1)$ random matrix $(T(\omega), h(\omega))$ by a certain sequence of random matrices

$$\big(T^1(\omega), h^1(\omega)\big), \big(T^2(\omega), h^2(\omega)\big), \ldots, \big(T^N(\omega), h^N(\omega)\big), \ldots,$$

converging in some probabilistic sense to $(T(\omega), h(\omega))$, we obtain the approximative objective functions

$$F^N(x) = \bar{c}'x + Ep\big(h^N(\omega) - T^N(\omega)x\big), \quad N = 1, 2, \ldots. \tag{6}$$

Using (4d), for $F(x) - F^N(x)$ we obtain, see [2],

$$-\vartheta^N(x) \leqslant F(x) - F^N(x) \leqslant \eta^N(x), \tag{7a}$$

where the lower, upper error term $\vartheta^N(x)$, $\eta^N(x)$, resp., is defined by

$$\vartheta^N(x) := Ep\big((T(\omega) - T^N(\omega))x + (h^N(\omega) - h(\omega))\big), \tag{7b}$$

$$\eta^N(x) := Ep\big((T^N(\omega) - T(\omega))x + (h(\omega) - h^N(\omega))\big). \tag{7c}$$

Using discretization methods, the approximations $(T^N(\omega), h^N(\omega))$ are piecewise constant random variables, hence

$$\big(T^N(\omega), h^N(\omega)\big) = \big(T^{N,j}, h^{N,j}\big) \quad \text{for all } \omega \in \Omega^{N,j},\ j = 1, 2, \ldots, r_N, \tag{8a}$$

where

$$\Omega^{N,1}, \Omega^{N,2}, \ldots, \Omega^{N,j}, \ldots, \Omega^{N,r_N} \text{ is a partition of } \Omega. \tag{8b}$$

Consequently, $F^N(x)$, given by (6), reads

$$F^N(x) = \overline{c}'x + \sum_{j=1}^{r_N} P(\Omega^{N,j})p\big(h^{N,j} - T^{N,j}x\big). \tag{8c}$$

Using Jensen's inequality, for the error estimate $\vartheta^N(x)$ we get

$$\begin{aligned}
\vartheta^N(x) &= \int p\big((T(\omega) - T^N(\omega))x + (h^N(\omega) - h(\omega))\big)P(d\omega) \\
&= \sum_{j=1}^{r_N} \int_{\Omega^{N,j}} p\big((T(\omega) - T^{N,j})x + (h^{N,j} - h(\omega))\big)P(d\omega) \\
&\geqslant \sum_{j=1}^{r_N} P(\Omega^{N,j})p\big((\overline{T}^{\Omega^{N,j}} - T^{N,j})x + (n^{N,j} - \overline{h}^{\Omega^{N,j}})\big),
\end{aligned} \tag{9a}$$

where

$$\overline{T}^{\Omega^{N,j}} = \frac{1}{P(\Omega^{N,j})} \int_{\Omega^{N,j}} T(\omega)P(d\omega), \tag{9b}$$

$$\overline{h}^{\Omega^{N,j}} = \frac{1}{P(\Omega^{N,j})} \int_{\Omega^{N,j}} h(\omega)P(d\omega) \tag{9c}$$

are the conditional expectation of $T(\omega)$, $h(\omega)$, resp., with respect to $\Omega^{N,j}$. Obviously, for $\eta^N(x)$ we obtain

$$\eta^N(x) \geqslant \sum_{j=1}^{r_N} P(\Omega^{N,j})p\big((T^{N,j} - \overline{T}^{\Omega^{N,j}})x + (\overline{h}^{\Omega^{N,j}} - h^{N,j})\big). \tag{9d}$$

In several publications [3–5] there is suggested to select the values $T^{N,j}$, $h^{N,j}$, resp., according to

$$T^{N,j} = \overline{T}^{\Omega^{N,j}} = \frac{1}{P(\Omega^{N,j})} \int\limits_{\Omega^{N,j}} T(\omega) P(\mathrm{d}\omega), \tag{10a}$$

$$h^{N,j} = \overline{h}^{\Omega^{N,j}} = \frac{1}{P(\Omega^{N,j})} \int\limits_{\Omega^{N,j}} h(\omega) P(\mathrm{d}\omega). \tag{10b}$$

In this case, (9a,d) and (4c) immediately yield this result:

LEMMA 2.1. *If the approximation* $(T^N(\omega), h^N(\omega))$ *of* $(T(\omega), h(\omega))$ *is defined by* (8a, b) *and* (10a, b), *then* $\vartheta^N(x) \geqslant 0$, $\eta^N(x) \geqslant 0$ *for all* $x \in \mathbb{R}^n$.

If the norm $\|p\|$ of the sublinear loss function is known, then the error terms $\vartheta^N(x), \eta^N(x)$ can be further estimated from above. Indeed, (7b, c) and (4f) yield

$$|\vartheta^N(x)|, |\eta^N(x)| \leqslant \|p\| \big(E\big\|\big(T(\omega) - T^N(\omega)\big)x\big\| + E\big\|h^N(\omega) - h(\omega)\big\|\big). \tag{11a}$$

Denoting by

$$P^{N,j}(\mathrm{d}\omega) = \frac{1}{P(\Omega^{N,j})} 1_{\Omega^{N,j}} P(\mathrm{d}\omega) \tag{11b}$$

the restriction of P to the subdomains $\Omega^{N,j}$ of Ω, according to (8a–c), from (11a, b) we obtain

$$\begin{aligned} &|\vartheta^N(x)|, |\eta^N(x)| \\ &\leqslant \|p\| \sum_{j=1}^{r_N} P(\Omega^{N,j}) \Big(\int_{\Omega^{N,j}} \big\|\big(T(\omega) - T^{N,j}\big)x\big\| P^{N,j}(\mathrm{d}\omega) + \\ &+ \int_{\Omega^{N,j}} \big\|h^{N,j} - h(\omega)\big\| P^{N,j}(\mathrm{d}\omega) \Big). \end{aligned} \tag{11c}$$

Remark 1 (More general loss functions). If the sublinear loss function p is replaced by a more general convex loss function u, then the inequalities (11a, c) remain true if the left-hand side in (11a, c) is simply replaced by $|F(x) - F^N(x)|$, and $\|p\|$ is replaced by a Lipschitz constant $L > 0$ of u, provided that u is Lipschitzian with constand L on the union of the supports of the random m-vectors $T(\omega)x - h(\omega)$ and $T^N(\omega)x - h^N(\omega)$, $N = 1, 2, \ldots, x \in D$, cf. [2].

Because of

$$\begin{aligned} \big\|\big(T(\omega) - T^{N,j}\big)x\big\| &= \big(\big\|\big(T(\omega) - T^{N,j}\big)x\big\|^2\big)^{1/2} \\ &= \big(x'\big(T(\omega) - T^{N,j}\big)'\big(T(\omega) - T^{N,j}\big)x\big)^{1/2} \end{aligned}$$

and the concavity of $t \to \sqrt{t}$, we have that

$$\int_{\Omega^{N,j}} \|(T(\omega) - T^{N,j})x\| P^{N,j}(d\omega)$$
$$\leqslant \left(\int_{\Omega^{N,j}} \|(T(\omega) - T^{N,j})x\|^2 P^{N,j}(d\omega) \right)^{1/2}$$
$$= \left(x' \left(\sum_{i=1}^{m} \int_{\Omega^{N,j}} (T_i(\omega) - T_i^{N,j})'(T_i(\omega) - T_i^{N,j}) P^{N,j}(d\omega) \right) x \right)^{1/2}$$
$$\leqslant \|x\| \left(\sum_{i=1}^{m} \sum_{k=1}^{n} \int_{\Omega^{N,j}} (t_{ik}(\omega) - t_{ik}^{N,j})^2 P^{N,j}(d\omega) \right)^{1/2}, \tag{12a}$$

where T_i, $T_i^{N,j}$ is the i-th row of the $m \times n$ matrices $T(\omega) = (t_{ik}(\omega))$, $T^{N,j} = (t_{ik}^{N,j})$, respectively. Moreover,

$$\int_{\Omega^{N,j}} \|h^{N,j} - h(\omega)\| P^{N,j}(d\omega)$$
$$\leqslant \left(\sum_{i=1}^{m} \int_{\Omega^{N,j}} (h_i^{N,j} - h_i(\omega))^2 P^{N,j}(d\omega) \right)^{1/2}. \tag{12b}$$

Clearly, if (10a, b) holds, then

$$\int_{\Omega^{N,j}} \|(T(\omega) - T^{N,j})x\| P^{N,j}(d\omega) \leqslant \left(x' \left(\sum_{i=1}^{m} \text{var}^{N,j}(T_i(\cdot)) \right) x \right)^{1/2}, \tag{12c}$$

$$\int_{\Omega^{N,j}} \|h^{N,j} - h(\omega)\| P^{N,j}(d\omega) \leqslant \left(\sum_{i=1}^{m} \text{var}^{N,j}(h_i(\cdot)) \right)^{1/2}, \tag{12d}$$

where $\text{var}^{N,j}(T_i(\cdot))$ is the covariance matrix of the i-th row $T_i(\omega)$ of $T(\omega)$, and $\text{var}^{N,j}(h_i(\cdot))$ is the variance of the i-th component $h_i(\omega)$ of $h(\omega)$ with respect to the conditional distribution $P^{N,j} = P|_{\Omega^{N,j}}$.

Remark 2. For the general case given by (12a, b) we have that

$$\int_{\Omega^{N,j}} (T_i(\omega) - T_i^{N,j})'(T_i(\omega) - T_i^{N,j}) P^{N,j}(d\omega)$$
$$= \text{var}^{N,j}(T_i(\cdot)) + (\overline{T}_i^{\Omega^{N,j}} - T_i^{N,j})'(\overline{T}^{\Omega^{N,j}} - T_i^{N,j}), \tag{12e}$$
$$\int_{\Omega^{N,j}} (h_i(\omega) - h_i^{N,j})^2 P^{N,j}(d\omega) = \text{var}^{N,j}(h_i(\cdot)) + (\overline{h}_i^{\Omega^{N,j}} - h_i^{N,j})^2, \tag{12f}$$

respectively.

Because of (11c) and (12a, b) we still have to compute an upper estimate of $\|p\|$. Representing any vector $z \in \mathbb{R}^m$ by $z = \sum_{i=1}^m z_i e_i$, where $z_1, z_2, \dots, z_m$ are the components of z and $e_1, e_2, \dots, e_n$ are the unit vectors of the m coordinate directions, because of the sublinearity of p we find

$$\begin{aligned} p(z) = p\left(\sum_{i=1}^m z_i e_i\right) &\leqslant \sum_{i=1}^m p(z_i e_i) = \sum_{i=1}^m p\big((z_i^+ - z_i^-)e_i\big) \\ &\leqslant \sum_{i=1}^m \big(p(z_i^+ e_i)\big) + p\big(z_i^-(-e_i)\big) = \sum_{i=1}^m \big(z_i^+ p(e_i) + z_i^- p(-e_i)\big) \\ &\leqslant \sum_{i=1}^m (z_i^+ + z_i^-) \max\big\{p(e_i), p(-e_i)\big\} = \sum_{i=1}^m \pi_i |z_i|, \end{aligned} \tag{13a}$$

where

$$\pi_i = \max\big\{p(e_i), p(-e_i)\big\}, \quad i = 1, 2, \dots, m. \tag{13b}$$

Note that for the computation of the m coefficients $\pi_1, \pi_2, \dots, \pi_m$ we have to solve the following $2m$ linear programs

$$\min\ q'y \quad \text{s.t. } Wy = \pm e_i, \quad y \geqslant 0 \text{ for } i = 1, 2, \dots, m. \tag{13c}$$

From (13a) we obtain

$$p(z) \leqslant \|\pi\| \cdot \|z\|, \tag{13d}$$

where $\|\pi\|, \|z\|$ denote the Euclidean norm of $\pi = (\pi_1, \pi_2, \dots, \pi_m)'$ and z, respectively. Thus, according to (4f) we have the upper norm bound

$$\|p\| \leqslant \|\pi\| = \left(\sum_{i=1}^m \pi_i^2\right)^{1/2} = \left(\sum_{i=1}^m \big(\max\{p(e_i), p(-e_i)\}\big)^2\right)^{1/2}, \tag{14}$$

where $\|\pi\|$ can be calculated from the given data, cf. (2c), (13c).

Summarizing the above consierations, from (11c), (12a–f), (14) we get:

THEOREM 2.1. *If* $(T^N(\omega), h^N(\omega))$ *is given by* (8a, b) *and* (3a, b) *holds, then for each* $x \in \mathbb{R}^n$ *we have that*

$$\begin{aligned} &|\vartheta^N(x)|, |\eta^N(x)| \\ &\leqslant \|\pi\| \sum_{j=1}^{r_N} P(\Omega^{N,j}) \left(\|x\| \left(\sum_{i,k=1}^m \int_{\Omega^{N,j}} \big(t_{ik}(\omega) - t_{ik}^{N,j}\big)^2 P^{N,j}(\mathrm{d}\omega) \right)^{1/2} + \right. \\ &\left. + \left(\sum_{i=1}^m \int_{\Omega_{N,j}} \big(h_i(\omega) - h_i^{N,j}\big)^2 P^{N,j}(\mathrm{d}\omega) \right)^{1/2} \right). \end{aligned} \tag{15a}$$

Moreover, if Equations (10a, b) *hold, then*

$$0 \leqslant \vartheta^N(x), \eta^N(x) \leqslant \|\pi\| \sum_{j=1}^{r_N} P(\Omega^{N,j}) \left(\|x\| \left(\sum_{i,k=1}^{m} \mathrm{var}^{N,j}(t_{ik}(\cdot)) \right)^{1/2} + \right.$$

$$\left. + \left(\sum_{i=1}^{m} \mathrm{var}^{N,j}(h_i(\cdot)) \right)^{1/2} \right). \tag{15b}$$

2.1. Special Representations of the Random Matrix $(T(\cdot), h(\cdot))$

A common, well-known representation of $(T(\omega), h(\omega))$ is given by

$$\big(T(\omega), h(\omega)\big) = \big(T(\xi(\omega)), h(\xi(\omega))\big) = (\overline{T}, \overline{h}) + \sum_{s=1}^{L} \xi_s(\omega)\big(T^{(s)}, h^{(s)}\big), \tag{16}$$

where $(\overline{T}, \overline{h})$ denotes the mean of $(T(\omega), h(\omega))$, $(T^{(s)}, h^{(s)})$, $s = 1, 2, \ldots, L$, are given $m \times (n+1)$ matrices, and $\xi_1(\omega), \xi_2(\omega), \ldots, \xi_L$, are zero-mean, stochastically independent random variables.

Based on representation (16), the approximation $(T^N(\omega), h^N(\omega))$ of $(T(\omega), h(\omega))$ can be described then according to (8a, b) by the following piecewise constant approximation $\xi^N(\omega)$ of the random L-vector $\xi(\omega) = (\xi_1(\omega), \xi_2(\omega), \ldots, \xi_L(\omega))$: Let Ξ denote the support of $\xi(\omega)$ or a set containing the support of $\xi(\omega)$. Moreover, let

$$\Xi^{N,1}, \Xi^{N,2}, \ldots, \Xi^{N,j}, \ldots, \Xi^{N,r_N} \tag{17a}$$

be the partition of Ξ generated by the partition (8b) of Ω, hence

$$\Xi^{N,j} = \big\{\xi(\omega) : \omega \in \Omega^{N,j}\big\}. \tag{17b}$$

Thus, with fixed L-vectors $\xi^{N,j} = (\xi_1^{N,j}, \ldots, \xi_L^{N,j}) \in \Xi^{N,j}$ we have that

$$\xi^N(\omega) = \xi^{N,j} \quad \text{for all } \omega \in \Omega^{N,j} \tag{17c}$$

and therefore

$$\big(T^{N,j}, h^{N,j}\big) = (\overline{T}, \overline{h}) + \sum_{s=1}^{L} \xi_s^{N,j}\big(T^{(s)}, h^{(s)}\big). \tag{17d}$$

Because of the properties of the random variables $\xi_1(\omega), \ldots, \xi_L(\omega)$ in the representation (16) of $(T(\omega), h(\omega))$, we suppose that $E\xi_s^N(\omega) = 0$ and $E\xi_s^N(\omega)\xi_t^N(\omega) = 0$ for all $s = 1, \ldots, L$, $t = 1, \ldots, L$, $t \neq s$, hence

$$\sum_{j=1}^{r_N} \xi_s^{N,j} P_{\xi(\cdot)}\left(\Xi^{N,j}\right) = 0, \qquad \sum_{j=1}^{r_N} \xi_s^{N,j} \xi_t^{N,j} P_{\xi(\cdot)}\left(\Xi^{N,j}\right) = 0,$$
$$s, t = 1, \ldots, L, t \neq s. \tag{17e}$$

In many cases we may suppose that

$$\Xi = \prod_{s=1}^{L} \Xi_s, \quad \Xi_s = [\alpha_s, \beta_s), \tag{18a}$$

is a half-open L-dimensional interval. In this case Ξ is then partitioned into certain subintervals $\Xi^{N,j}$. Hence, we set

$$j = (j_1, j_2, \ldots, j_L), \qquad r_N = (r_{N1}, r_{N2}, \ldots, r_{NL}), \tag{18b}$$
$$\xi^{N,j} = \left(\xi_1^{N,j_1}, \xi_2^{N,j_2}, \ldots, \xi_L^{N,j_L}\right), \tag{18c}$$

where $j_s = 1, 2, \ldots, r_{N^s}, s = 1, \ldots, L$, and a cell $\Xi^{N,j}$ is given by

$$\Xi^{N,j} = \Xi^{N,(j_1,\ldots,j_L)} = \prod_{s=1}^{L} \Xi_s^{N,j_s} \tag{18d}$$

with certain half-open subintervals

$$\Xi_s^{N,j_s} = \left[\alpha_s^{N,j_s}, \beta_s^{N,j_s}\right), \quad j_s = 1, 2, \ldots, r_{N^s}, s = 1, 2, \ldots, L, \tag{18e}$$

and

$$\xi_s^{N,j_s} \in \left[\alpha_s^{N,j_s}, \beta_s^{N,j_s}\right), \quad j_s = 1, 2, \ldots, r_{N^s}, s = 1, 2, \ldots, L. \tag{18f}$$

Moreover, (17d) reads in the present case

$$\begin{aligned}\left(T^{N,j}, h^{N,j}\right) &= \left(T^{N,(j_1,\ldots,j_L)}, h^{N,(j_1,\ldots,j_L)}\right) \\ &= (\overline{T}, \overline{h}) + \sum_{s=1}^{L} \xi_s^{N,j_s}\left(T^{(s)}, h^{(s)}\right).\end{aligned} \tag{18g}$$

For the integrals in the error estimation (15a), by (16), (17d), (18g) and the cell representation (18d), we get

$$\int_{\Omega^{N,j}} \left(t_{ik}(\omega) - t_{ik}^{N,j}\right)^2 P^{N,j}(\mathrm{d}\omega) = \frac{1}{P_{\xi(\cdot)}(\Xi^{N,j})} \int_{\xi \in \Xi^{N,j}} \left(t_{ik}(\xi) - t_{ik}^{N,j}\right)^2 P_{\xi(\cdot)}(\mathrm{d}\xi)$$

and therefore

$$\int_{\Omega^{N,j}} \left(t_{ik}(\omega) - t_{ik}^{N,j}\right)^2 P^{N,j}(\mathrm{d}\omega)$$

$$= \sum_{\substack{s,\sigma=1 \\ s\neq\sigma}}^{L} t_{ik}^{(s)} t_{ik}^{(\sigma)} \left(\bar{\xi}_s^{\Xi_s^{N,j_s}} - \xi_s^{N,j_s}\right)\left(\bar{\xi}_\sigma^{\Xi_\sigma^{N,j_\sigma}} - \xi_\sigma^{N,j_\sigma}\right) +$$

$$+ \sum_{s=1}^{L} t_{ik}^{(s)2} \frac{1}{P_{\xi_s(\cdot)}(\Xi_s^{N,j_s})} \int_{\xi_s \in \Xi_s^{N,j_s}} \left(\xi_s - \xi_s^{N,j_s}\right)^2 P_{\xi_s(\cdot)}(d\xi_s) \quad (19a)$$

as well as

$$\int_{\Omega^{N,j}} \left(h_i(\omega) - h_i^{N,j}\right)^2 P^{N,j}(d\omega)$$

$$= \sum_{\substack{s,\sigma=1 \\ s\neq\sigma}}^{L} h_i^{(s)} h_i^{(\sigma)} \left(\bar{\xi}_s^{\Xi_s^{N,j_s}} - \xi_s^{N,j_s}\right)\left(\bar{\xi}_\sigma^{\Xi_\sigma^{N,j_\sigma}} - \xi_\sigma^{N,j_\sigma}\right) +$$

$$+ \sum_{s=1}^{L} h_i^{(s)2} \frac{1}{P_{\xi_s(\cdot)}(\Xi_s^{N,j_s})} \int_{\xi_s \in \Xi_s^{N,j_s}} \left(\xi_s - \xi_s^{N,j_s}\right)^2 P_{\xi_s(\cdot)}(d\xi_s), \quad (19b)$$

where

$$\bar{\xi}_s^{\Xi_s^{N,j_s}} = E\left(\xi_s \mid \Xi_s^{N,j_s}\right) := \frac{1}{P_{\xi_s(\cdot)}(\Xi_s^{N,j_s})} \int_{\xi_s \in \Xi_s^{N,j_s}} \xi_s P_{\xi_s(\cdot)}(d\xi_s) \quad (19c)$$

is the conditional mean of $\xi_s(\omega)$ with respect to Ξ_s^{N,j_s}, cf. (9b, c).

Since $\xi_s^{N,j_s} \in [\alpha_s^{N,j_s}, \beta_s^{N,j_s})$, cf. (18f), and $\bar{\xi}_s^{\Xi_s^{N,j_s}} \in [\alpha_s^{N,j_s}, \beta_s^{N,j_s})$, the first terms in (19a, b) can be estimated from above as follows:

$$\left| \sum_{\substack{s,\sigma=1 \\ s\neq\sigma}}^{L} t_{ik}^{(s)} t_{ik}^{(\sigma)} \left(\bar{\xi}_s^{\Xi_s^{N,j_s}} - \xi_s^{N,j_s}\right)\left(\bar{\xi}_\sigma^{\Xi_\sigma^{N,j_\sigma}} - \xi_\sigma^{N,j_\sigma}\right) \right|$$

$$\leqslant \sum_{\substack{s,\sigma=1 \\ s\neq\sigma}}^{L} \left| t_{ik}^{(s)} t_{ik}^{(\sigma)} \right| \left(\beta_s^{N,j_s} - \alpha_s^{N,j_s}\right)\left(\beta_\sigma^{N,j_\sigma} - \alpha_\sigma^{N,j_\sigma}\right), \quad (19d)$$

$$\left| \sum_{s,\sigma=1}^{L} h_i^{(s)} h_i^{(\sigma)} \left(\bar{\xi}_s^{\Xi_s^{N,j_s}} - \xi_s^{N,j_s}\right)\left(\bar{\xi}_\sigma^{\Xi_\sigma^{N,j_\sigma}} - \xi_\sigma^{N,j_\sigma}\right) \right|$$

$$\leqslant \sum_{\substack{s,\sigma=1 \\ s\neq\sigma}}^{L} \left| h_i^{(s)} h_i^{(\sigma)} \right| \left(\beta_s^{N,j_s} - \alpha_s^{N,j_s}\right)\left(\beta_\sigma^{N,j_\sigma} - \alpha_\sigma^{N,j_\sigma}\right). \quad (19e)$$

If the interval $\Xi_s = [\alpha_s, \beta_s)$ is partitioned into r_{Ns} equidistant subintervals $[\alpha_s^{N,j_s}, \beta_s^{N,j_s})$, $j_s = 1, 2, \ldots, r_{Ns}$, then

$$\sum_{\substack{s,\sigma=1\\ s\neq\sigma}}^{L} \left|t_{ik}^{(s)} t_{ik}^{(\sigma)}\right| \left(\beta_s^{N,j_s} - \alpha_s^{N,j_s}\right)\left(\beta_\sigma^{N,j_\sigma} - \alpha_\sigma^{N,j_\sigma}\right)$$

$$= \frac{(\beta_s - \alpha_s)}{r_{Ns}} \frac{(\beta_\sigma - \alpha_\sigma)}{r_{N\sigma}} \sum_{\substack{s,\sigma=1\\ s\neq\sigma}}^{L} \left|t_{ik}^{(s)} t_{ik}^{(\sigma)}\right|, \tag{19f}$$

$$\sum_{\substack{s,\sigma=1\\ s\neq\sigma}}^{L} \left|h_i^{(s)} h_i^{(\sigma)}\right| \left(\beta_s^{N,j_s} - \alpha_s^{N,j_s}\right)\left(\beta_\sigma^{N,j_\sigma} - \alpha_\sigma^{N,j_\sigma}\right)$$

$$= \frac{(\beta_s - \alpha_s)^2}{r_{Ns} r_{N\sigma}} \sum_{\substack{s,\sigma=1\\ s\neq\sigma}}^{L} \left|h_i^{(s)} h_i^{(\sigma)}\right|. \tag{19g}$$

If the values ξ_s^{N,j_s}, $j_s = 1, 2, \ldots, r_{Ns}$, $s = 1, 2, \ldots, L$, are selected such that

$$\xi_s^{N,j_s} = \bar{\xi}_s^{\Xi_s^{N,j_s}}, \quad j_s = 1, 2, \ldots, r_{NS}, s = 1, 2, \ldots, L, \tag{20a}$$

see (10a, b), then (19a, b) is reduced to

$$\int_{\Omega^{N,j}} \left(t_{ik}(\omega) - t_{ik}^{N,j}\right)^2 P^{N,j}(\mathrm{d}\omega) = \sum_{s=1}^{L} t_{ik}^{(s)2} \mathrm{var}^{N,j_s}(\xi_s(\cdot)), \tag{20b}$$

$$\int_{\Omega^{N,j}} \left(h_i(\omega) - h_i^{N,j}\right)^2 P^{N,j}(\mathrm{d}\omega) = \sum_{s=1}^{L} h_i^{(s)2} \mathrm{var}^{N,j_s}(\xi_s(\cdot)), \tag{20c}$$

where

$$\mathrm{var}^{N,j_s}(\xi_s(\cdot)) = \frac{1}{P_{\xi_s(\cdot)}(\Xi_s^{N,j_s})} \int_{\xi_s \in \Xi_s^{N,j_s}} \left(\xi_s - \bar{\xi}_s^{\Xi_s^{N,j_s}}\right)^2 P_{\xi_s(\cdot)}(\mathrm{d}\xi_s) \tag{20d}$$

is the conditional variance of $\xi_s(\omega)$ with respect to Ξ_s^{N,j_s}.

EXAMPLE 2.1. Suppose that $\xi_s(\omega)$ has a density $f_s(z)$ such that for all $s = 1, 2, \ldots, L$ and $j_s = 1, 2, \ldots, r_{N,s}$ we have that

$$\begin{aligned} 0 < f_{s,m}^{N,j_s} &:= \inf\left\{f_s(z) : z \in \Xi_s^{N,j_s}\right\} \leqslant f_{s,M}^{N,j_s} \\ &:= \sup\left\{f_s(z) : z \in \Xi_s^{N,j_s}\right\} < +\infty. \end{aligned} \tag{21a}$$

This yields

$$P_{\xi_s(\cdot)}\left(\Xi_s^{N,j_s}\right) \geqslant f_{s,m}^{N,j_s}\left(\beta_s^{N,j_s} - \alpha_s^{N,j_s}\right), \tag{21b}$$

$$\int_{\xi_s \in \Xi_s^{N,j_s}} \left(\xi_s - \overline{\xi}_s^{\Xi_s^{N,j_s}}\right)^2 P_{\xi_s(\cdot)}(\mathrm{d}\xi_s) \leqslant f_{s,M}^{N,j_s} \int_{\alpha_s^{N,j_s}}^{\beta_s^{N,j_s}} \left(\xi_s - \overline{\xi}_s^{\Xi_s^{N,j_s}}\right)^2 \mathrm{d}\xi_s \tag{21c}$$

and therefore

$$\begin{aligned} \mathrm{var}^{N,j_s}(\xi_s(\cdot)) &\leqslant \frac{f_{s,M}^{N,j_s}}{f_{s,m}^{N,j_s}} \left(\frac{1}{\beta_s^{N,j_s} - \alpha_s^{N,j_s}} \int_{\alpha_s^{N,j_s}}^{\beta_s^{N,j_s}} \left(\xi_s - \frac{\alpha_s^{N,j_s} + \beta_s^{N,j_s}}{2}\right)^2 \mathrm{d}\xi_s + \right. \\ &\quad \left. + \left(\frac{\alpha_s^{N,j_s} + \beta_s^{N,j_s}}{2} - \overline{\xi}_s^{\Xi_s^{N,j_s}} \right)^2 \right) \\ &= \frac{f_{s,M}^{N,j_s}}{f_{s,m}^{N,j_s}} \left(\frac{(\beta_s^{N,j_s} - \alpha_s^{N,j_s})^2}{12} + \left(\frac{\alpha_s^{N,j_s} + \beta_s^{N,j_s}}{2} - \overline{\xi}_s^{\Xi_s^{N,j_s}} \right)^2 \right). \end{aligned} \tag{21d}$$

Since $\frac{\alpha_s^{N,j_s} + \beta_s^{N,j_s}}{2}$ and $\overline{\xi}_s^{\Xi_s^{N,j_s}}$ are elements of $\Xi_s^{N,j_s} = [\alpha_s^{N,j_s}, \beta_s^{N,j_s})$, from (20d) and (21a–d) we obtain

$$\mathrm{var}^{N,j_s}(\xi_s(\cdot)) \leqslant \frac{f_{s,M}^{N,j_s}}{f_{s,m}^{N,j_s}} \frac{13}{12} \left(\beta_s^{N,j_s} - \alpha_s^{N,j_s}\right)^2. \tag{22a}$$

If each Ξ_s is partitioned into r_{Ns} equidistant subintervals Ξ_s^{N,j_s}, $j_s = 1, 2, \ldots, r_{Ns}$, then $\beta_s^{N,j_s} - \alpha_s^{N,j_s} = \dfrac{1}{r_{Ns}}$ for all $j = 1, 2, \ldots, r_{Ns}$ and therefore

$$\mathrm{var}^{N,j_s}(\xi_s(\cdot)) \leqslant \frac{13}{12} \frac{f_{s,M}^{N,j_s}}{f_{s,m}^{N,j_s}} \cdot \frac{1}{r_{Ns}^2}. \tag{22b}$$

3. Approximations of F with a Given Error Level ε

According to (7a–c) and (15a, b) we have that

$$\left|F(x) - F^N(x)\right| \leqslant \|\pi\| \sum_{j=1}^{r_N} P\left(\Omega^{N,j}\right)\left(\|x\| V^{N,j}(T(\cdot)) + V^{N,j}(h(\cdot))\right), \tag{23a}$$

with the estimation errors

$$V^{N,j}(T(\cdot)) = \left(\sum_{i,k=1}^{N} \int_{\Omega^{N,j}} \left(t_{ik}(\omega) - t_{ik}^{N,j}\right)^2 P^{N,j}(\mathrm{d}\omega) \right)^{1/2}, \tag{23b}$$

$$V^{N,j}(h(\cdot)) = \left(\sum_{i,k=1}^{N} \int_{\Omega^{N,j}} \left(h_i(\omega) - h_i^{N,j}\right)^2 P^{N,j}(\mathrm{d}\omega) \right)^{1/2}. \tag{23c}$$

Knowing that D is bounded, hence

$$D \subset \{x \in \mathbb{R}^n : \|x\| \leqslant \rho_0\} \tag{24a}$$

for some $\rho_0 > 0$, from (23a) we obviously get

$$|F^* - F^{N*}| \leqslant \|\pi\| \sum_{j=1}^{r_N} P(\Omega^{N,j})(\rho_0 V^{N,j}(T(\cdot)) + V^{N,j}(h(\cdot))), \tag{24b}$$

where F^* is the optimal value of (2a–c) and F^{N*} the optimal value of the approximating problem

$$\min F^N(x) \quad \text{s.t. } x \in D. \tag{24c}$$

Furthermore, if it is known that there is an optimal solution x^* of (2a–c) such that with some $p \geqslant 1$ we have that

$$\|x^*\|_p \leqslant \rho_{0p} \quad \text{for some given } \rho_{0p} > 0, \tag{25a}$$

where $\|x\|_p$ is the p-norm of x, then again (7a–c) and (15a, b) yield

$$|F^* - F^{N*}_{\rho_0}| \leqslant \|\pi\| \sum_{j=1}^{r_N} P(\Omega^{N,j})(\rho_0 V^{N,j}(T(\cdot)) + V^{N,j}(h(\cdot))), \tag{25b}$$

where $\rho_0 := \rho_{0p} \max\{\|u\| : \|u\|_p \leqslant 1\}$, and $F^{N*}_{\rho_0}$ is the optimal value of the approximation

$$\min F^N(x) \quad \text{s.t. } x \in D, \|x\|_p \leqslant \rho_{0p}. \tag{25c}$$

Note that if $F^N(x)$ is generated by a discretization process of $P_{(T(\cdot),h(\cdot))}$, and $p = 1$ or $p = 1$ or $p = +\infty$, then (25c) can again be represented by a *linear program.*

The above considerations yield now the following result:

THEOREM 3.1. *Selecting the approximation* $(T^N(\omega), h^N(\omega))$ *of* $(T(\omega), h(\omega))$ *such that*

$$\|\pi\| \sum_{j=1}^{r_N} P(\Omega^{N,j})(\rho_0 V^{N,j}(T(\cdot)) + V^{N,j}(h(\cdot))) \leqslant \varepsilon, \tag{26a}$$

where $\varepsilon > 0$ *is a given error bound, then in cases* (24a) *and* (25a) *we have the* a priori error bound

$$|F^* - F^{N*}| < \varepsilon, \qquad |F^* - F^{N*}_{\rho_0}| < \varepsilon, \tag{26b}$$

respectively.

While (24a) is a simple property of the convex polyhedron D which may hold in some cases, the relation (25a) is more involved.

4. Norm Bounds for Optimal Solutions of (2a–c)

For finding upper norm bounds ρ_0 for an optimal solution x^* of (2a–c) we first have to study the growth properties of F. These can be obtained if for the loss function p, see (2c), appropriate lower bounds can be derived.

Using condition (3a, b), where we assume that q has components

$$q_1 > 0, \quad q_2 > 0, \quad \ldots, \quad q_\mu > 0, \tag{27a}$$

we define now the closed, convex polyhedron K by

$$K = \operatorname{conv}\left\{\frac{1}{q_1}w_1, \frac{1}{q_2}w_2, \ldots, \frac{1}{q_\mu}w_\mu\right\}, \tag{27b}$$

where $q_k > 0, k = 1, \ldots, \mu$, are the components of q and $w_1, w_2, \ldots, w_\mu$ are the columns of W, where we may assume – without any restrictions – that $w_k \neq 0$ for all $k = 1, \ldots, \mu$. According to [2,6] we know that in this situation the loss function p has the representation

$$p(z) = \inf\left\{\lambda > 0 : \frac{z}{\lambda} \in K\right\}, \quad z \in \mathbb{R}^m. \tag{28}$$

Having (27a, b) and defining

$$q_0 = \max_{1 \leqslant k \leqslant \mu} \frac{1}{q_k}\|w_k\|, \tag{29a}$$

where $q_0 > 0$, we find

$$\|z\| \leqslant q_0 \quad \text{for each } z \in K.$$

Since the relation $\frac{1}{\lambda}z \in K$ obviously implies that $\|\frac{1}{\lambda}z\| \leqslant q_0$, for the loss function p, having representation (28), for each $z \in \mathbb{R}^m$ we have that

$$p(z) \geqslant \inf\left\{\lambda > 0 : \left\|\frac{z}{\lambda}\right\| \leqslant q_0\right\} = \frac{1}{q_0}\|z\| = \left(\min_{1 \leqslant k \leqslant \mu} \frac{q_k}{\|w_k\|}\right)\|z\|. \tag{29b}$$

If (29b) holds, then (5) yields

$$F(x) \geqslant \bar{c}'x + \underline{p}E\|T(\omega)x - h(\omega)\|, \tag{30a}$$

where

$$\underline{p} = \min_{1 \leqslant k \leqslant \mu} \frac{q_k}{\|w_k\|}. \tag{30b}$$

In the following we assume that $(T(\omega), h(\omega))$ is bounded w.p.1, hence, there is a constant $\Gamma > 0$ such that

$$\left\|\left(T(\omega), h(\omega)\right)\right\| \leqslant \Gamma \quad \text{w.p.1.} \tag{31}$$

Defining for any $x \in \mathbb{R}^n$

$$\hat{x} = (x', 1)', \quad e_{\hat{x}} = \frac{\hat{x}}{\|\hat{x}\|},$$

we find

$$\|T(\omega)x - h(\omega)\| = \|(T(\omega), h(\omega))\hat{x}\| = \|\hat{x}\| \cdot \|(T(\omega), h(\omega)e_{\hat{x}}\|;$$

furthermore, we have that

$$\begin{aligned} \|(T(\omega), h(\omega))e_{\hat{x}}\|^2 &= \|(T(\omega), h(\omega))e_{\hat{x}}\| \cdot \|(T(\omega), h(\omega))e_{\hat{x}}\| \\ &\leqslant \|(T(\omega), h(\omega))e_{\hat{x}}\| \cdot \Gamma \end{aligned}$$

and therefore

$$\|T(\omega)x - h(\omega)\| \geqslant \frac{1}{\Gamma}\|\hat{x}\| \cdot \|(T(\omega), h(\omega))e_{\hat{x}}\|^2, \tag{32a}$$

see (31). Taking expectations on both sides of (32a), we get

$$E\|T(\omega)x - h(\omega)\| \geqslant \frac{1}{\Gamma}\|\hat{x}\| e'_{\hat{x}} E(T(\omega), h(\omega))'(T(\omega), h(\omega))e_{\hat{x}}. \tag{32b}$$

Denoting by $\lambda_{\min}(Q)$ the minimal eigenvalue of any symmetric matrix Q, from (32b) we obtain

$$\begin{aligned} E\|T(\omega)x - h(\omega)\| &\geqslant \frac{1}{\Gamma}\|\hat{x}\| \lambda_{\min}\big(E(T(\omega), h(\omega))'(T(\omega), h(\omega))\big) \\ &\geqslant \frac{1}{\Gamma}\|\hat{x}\| E\lambda_{\min}\big((T(\omega), h(\omega))'(T(\omega), h(\omega))\big). \end{aligned} \tag{32c}$$

Note that

$$\begin{aligned} &E(T(\omega), h(\omega))'(T(\omega), h(\omega)) \\ &\quad = \sum_{i=1}^{m} \big(\mathrm{var}(T_i(\cdot), h_i(\cdot)) + (\overline{T}_i, \overline{h}_i)'(\overline{T}_i, \overline{h}_i)\big), \end{aligned} \tag{32d}$$

where $\mathrm{var}(T_i(\cdot), h_i(\cdot))$ designates the covariance matrix of the i-th row (T_i, h_i) of $(T(\omega), h(\omega))$. If the random matrix $(T(\omega), h(\omega))$ is represented by (16), then

$$E(T(\omega), h(\omega))'(T(\omega), h(\omega)) = \overline{T}'T + \sum_{s=1}^{L} \mathrm{var}\,(\xi_s) T^{(s)'} T^{(s)}, \tag{32e}$$

hence, this matrix can be computed easily. Let then λ_0 be defined by

$$\begin{aligned} &\lambda_0 := \lambda_{\min}\big(E(T(\omega), h(\omega))'(T(\omega), h(\omega))\big) \quad \text{or} \\ &\lambda_0 = E\lambda_{\min}(T(\omega), h(\omega))'(T(\omega), h(\omega)). \end{aligned} \tag{33}$$

Summarizing the above considerations, according to (30a) and (32b) we find the following result:

LEMMA 4.1. *Suppose that conditions* (3a,b) *and* (31) *hold true. If* $\underline{p}, \Gamma, \lambda_0$ *are defined by* (30b), (31), (33), *resp., then*

$$F(x) \geqslant \overline{c}'x + \frac{\underline{p}\lambda_0}{\Gamma}\|\hat{x}\| \quad \text{for all } x \in \mathbb{R}^n. \tag{34}$$

Consider now any element x^0 of the feasible domain D of (2a). If x^* is an optimal solution of (2a–c), then (34) and (13d) yield the following inequalities

$$\begin{aligned}
&\overline{c}'x^* + \frac{\underline{p}\lambda_0}{\Gamma}\|\widehat{x^*}\| \\
&\leqslant F(x^*) \leqslant F(x^0) \\
&= \overline{c}'x^0 + \|\pi\| E \big\| T(\omega)x^0 - h(\omega) \big\| \\
&\leqslant F^0 := \overline{c}'x^0 + \|\pi\| \big(\widehat{x^0} E\big(T(\omega), h(\omega)\big)'\big(T(\omega), h(\omega)\big)\widehat{x^0}\big)^{1/2},
\end{aligned} \tag{35}$$

where the last inequality is guaranteed by the concavity of the function $z \rightarrow \sqrt{z}$. Note that the upper bound F^0 can be computed easily, see (14), (32e). Since $\|\widehat{x^*}\| = (1 + \|x^*\|^2)^{1/2}$, we have now the following norm bounds for optimal solutions x^* of (2a–c).

THEOREM 4.1. *Suppose that the assumptions of Lemma* 4.1 *hold, and let* x^0 *be any feasible solution of* (2a–c). *Moreover, let* F^0 *be defined as in* (35).

(a) *If* $\underline{c}'x \geqslant 0$ *for all* $x \in D$, *then for any optimal solution* x^* *of* (2a–c) *we have that*

$$\|x^*\| \leqslant \left(\left(\frac{\Gamma F^0}{\underline{p}\lambda_0}\right)^2 - 1\right)^{1/2}. \tag{36a}$$

(b) *If* $\|\overline{c}\| < \frac{\underline{p}\lambda_0}{\Gamma}$, *then for any optimal solution* x^* *of* (2) *it holds*

$$\begin{aligned}
\|x^*\| &\leqslant \left(\left(\frac{F(x^0)\Gamma}{\underline{p}\lambda_0 - \Gamma\|\overline{c}\|}\right)^2 - 1\right)^{1/2} \\
&\leqslant \left(\left(\frac{F^0\Gamma}{\underline{p}\lambda_0 - \Gamma\|\overline{c}\|}\right)^2 - 1\right)^{1/2}.
\end{aligned} \tag{36b}$$

PROOF. (a) Here, from (34) and (35) we get

$$F^0 \geqslant F(x^0) \geqslant \overline{c}'x^* + \frac{\underline{p}\lambda_0}{\Gamma}\|\widehat{x^*}\| \geqslant \frac{\underline{p}\lambda_0}{\Gamma}\left(1 + \|x^*\|^2\right)^{1/2}$$

which yields the first assertion (36a).

(b) The next assertion (36b) follows from

$$F^0 \geqslant F(x^0) \geqslant \overline{c}'x^* + \frac{p\lambda_0}{\Gamma}\|\widehat{x^*}\| \geqslant -\|\overline{c}\| \cdot \|x^*\| + \frac{p\lambda_0}{\Gamma}\|\widehat{x^*}\|$$
$$\geqslant \left(-\|\overline{c}\| + \frac{p\lambda_0}{\Gamma}\right)(1 + \|x^*\|^2)^{1/2}.$$

5. Invariant Discretizations

According to Theorem 2.1, (23a–c), there is a large variety of possible discretizations (T^N, h^N) of $(T(\omega), h(\omega))$ guaranteeing a certain given a priori error bound, see (26b). Hence, the problem is to find discretizations taking into consideration the special structure of the underlying problem [7]. A main idea in stochastic linear programming with recourse is the use of special refining strategies for refining the partitions $\Xi^{N,1}, \ldots, \Xi^{N,r_N}$ of Ξ, see (17a–e), such that only cells $\Xi^{N,j}$ are further partitioned which contribute most to the increase of the accuracy of approximation, see [3,4].

Very often the probability distribution $P_{(T(\cdot),h(\cdot))}$ has certain symmetry or invariance properties, cf. [8–11]. Not destroying these invariance properties during the discretization process, in several cases descent discretions can be constructed very easily.

Considering the approximation $(T^N(\omega), h^N(\omega))$, given by (8a, b) or (17a–e), we define $(T_0^N(\omega), h_0^N(\omega))$ by

$$\left(T_0^N(\omega), h_0^N(\omega)\right) := \left(T^N(\omega) - \overline{T}^N, h^N(\omega) - \overline{h}^N\right), \tag{37a}$$

where $(\overline{T}^N, \overline{h}^N)$ is the mean of $(T^N(\omega), h^N(\omega))$. Using the results of [11], we define the distribution invariance as follows, where the set $\mathcal{B}_\alpha$ of $r_N \times r_N$ matrices $B = (b_{ij})$ is given by

$$\mathcal{B}_\alpha = \{B : 1'B = 1', B\alpha = \alpha, B \geqslant 0\}. \tag{37b}$$

Here, 1 denotes the r_N-vector $1 = (1, 1, \ldots, 1)$, α is the r_N-vector

$$\alpha = \left(P(\Omega^{N,1}), P(\Omega^{N,2}), \ldots, P(\Omega^{N,r_N})\right)' \tag{37c}$$

or

$$\alpha = \left(P_{\xi(\cdot)}(\Xi^{N,1}), P_{\xi(\cdot)}(\Xi^{N,2}), \ldots, P_{\xi(\cdot)}(\Xi^{N,r_N})\right)' \tag{37d}$$

and $B \geqslant 0$ means that $b_{ij} \geqslant 0$ for all elements b_{ij} of B.

Definition 5.1. The distribution $P_{(T_0^N(\cdot),h_0^N(\cdot))}$ of $(T_0^N(\omega), h_0^N(\omega))$ is called *invariant* if there is a matrix $B \in \mathcal{B}_\alpha$ and an $n \times n$ matrix C such that for each row $i = 1, 2, \ldots, m$ we have that

$$B' \begin{pmatrix} T_{0,i}^{N,1} \\ T_{0,i}^{N,2} \\ \vdots \\ T_{0,i}^{N,r_N} \end{pmatrix} = \begin{pmatrix} T_{0,i}^{N,1} \\ T_{0,i}^{N,2} \\ \vdots \\ T_{0,i}^{N,r_N} \end{pmatrix} C, \tag{38a}$$

$$B' \begin{pmatrix} h_{0,i}^{N,1} \\ h_{0,i}^{N,2} \\ \vdots \\ T_{0,i}^{N,r_N} \end{pmatrix} = \begin{pmatrix} h_{0,i}^{N,1} \\ h_{0,i}^{N,2} \\ \vdots \\ h_{0,i}^{N,r_N} \end{pmatrix}. \tag{38b}$$

For the general case, we have to introduce some more notations:
Let denote $\tilde{z}$ the $(1+m)$-vector

$$\tilde{z} = \begin{pmatrix} t \\ z \end{pmatrix} \quad \text{with } t \in \mathbb{R}, z \in \mathbb{R}^m, \tag{39a}$$

where we set $\tilde{z} = (\tilde{z}_0, \tilde{z}_1, \ldots, \tilde{z}_m)$; with $\tilde{z}_0 = t, \tilde{z}_i = z_i, i = 1, \ldots, m_i$; furthermore, let $u(\tilde{z})$ denote the total loss function

$$u(\tilde{z}) = t + p(z) \tag{39b}$$

of (2a–c).

Obviously, the total loss function $u = u(\tilde{z})$ is monotoneous nondecreasing with respect to the component $z_0 = t$. In many cases the loss function p itself has some (partial) monotonicity properties, see, e.g., [10]. Hence, supposing in the following – for example – that p is also *partically nondecreasing*, we have a subset $J \subset \{0, 1, \ldots, m\}$ with $0 \in J$ and a corresponding partition

$$\tilde{z} = \begin{pmatrix} \tilde{z}_I \\ \tilde{z}_{II} \end{pmatrix}, \quad \tilde{z}_I = (\tilde{z}_i)_{i \in J}, \tilde{z}_{II} = (z_i)_{i \notin J}, \tag{40a}$$

of $\tilde{z}$ into two subvectors $\tilde{z}_I, \tilde{z}_{II}$ such that for any vectors $\tilde{z}, \tilde{w} \in \mathbb{R}^{1+m}$ the following relation hold:

$$\tilde{z}_I \leqslant \tilde{w}_I, \qquad \tilde{z}_{II} = \tilde{w}_{II} \Rightarrow u(\tilde{z}) \leqslant u(\tilde{w}), \tag{40b}$$

where $\tilde{z}_I \leqslant \tilde{w}_I$ means that $z_i \leqslant w_i$ for all $i \in J$. Of course, in many cases we have also this sharper condition

$$\tilde{z}_I \leqslant \tilde{w}_I, \qquad \tilde{z}_{II} = \tilde{w}_{II}, \qquad \tilde{z}_i < \tilde{w}_i$$
$$\text{for at least one } i \in J \Rightarrow u(\tilde{z}) < u(\tilde{w}). \tag{40c}$$

Based on the above definitions, the invariance of an arbitrary distribution $P_{(A^N(\cdot), b^N(\cdot))}$ with

$$(A^N(\omega), b^N(\omega)) = \begin{pmatrix} \overline{c}' & 0 \\ T^N(\omega) & h^N(\omega) \end{pmatrix} \tag{41}$$

is stated as follows, where the following inclusion is still assumed:

$$D \subset \mathbb{R}^n_+. \tag{42}$$

DEFINITION 5.2. The probability distribution $P_{(A^N(\cdot),b^N(\cdot))}$ of $(A^N(\omega), b^N(\omega))$ is called *invariant* it there is a matrix $B \in \mathcal{B}_\alpha$ and an $n \times n$ matrix C such that the following relations hold:

(i)

$$\bar{c} \geqslant \bar{c}'C, \tag{43a}$$

(ii)

$$\overline{T}_I \leqslant \overline{T}_I C, \tag{43b}$$

$$\overline{T}_{II} = \overline{T}_{II} C, \tag{43c}$$

(iii)

(38a) and (38b) are fulfilled, (43d)

where $\overline{T}_I, \overline{T}_{II}$, resp. is the matrix containing the rows $\overline{T}_i$ with $i \in J, i \notin J$, respectively.

Remark 5.1. Note that condition (43d), hence, relations (38a) and (38b) can be interpreted as *conditions for the discretization* of the distribution of the centralized random matrix $(T_0(\omega), h_0(\omega)) = (T(\omega) - \overline{T}, h(\omega) - \bar{h})$, where $(\overline{T}, \bar{h})$ is the mean of $(T(\omega), h(\omega))$.

The significance of the above invariance concept follows from the following result, cf. [11].

THEOREM 5.1. *Suppose that $D \subset \mathbb{R}^n_+$. If $(A^N(\cdot), b^N(\cdot))$ has an invariant distribution with matrices $B \in \mathcal{B}_\alpha$, C according to Definition 5.2, then*

(a) $F^N(y) \leqslant F^N(x)$ *with* $y := Cx$ *for every* $x \in \mathbb{R}^n$,

(b) $h = y - x$ *is a descent direction for* F^N at x, *provided that only* F^N *is not constant on the line segment* xy *joining* x *and* $y \neq x$.

As an important consequence of Theorem 5.1 we find the following result:

COROLLARY 5.1. *Assume that $P_{(A^N(\cdot),b^N(\cdot))}$ is invariant with matrices $B \in \mathcal{B}_\alpha, C$ according to Definition 5.2. Furthermore, suppose that $F^N(x)$ is not constant on each line segment xy in D. If x^* is an optimal solution of the approximating problem* (24c), *then*

$$Cx^* = x^* \quad or$$
$$h = Cx^* - x^* \quad \text{is not a feasible direction for } D \text{ at } x^*.$$

Note. Corollary 5.1 holds also under weakes conditions concerning F^N.

Bibliography

[1] Marti, K.: Optimal design of trusses as a stochastic linear programming problem, In: A. S. Nowak (ed.), *Reliability and Optimization of Structural Systems*, University of Michigan Press, Ann Arbor, 1998, pp. 231–239.

[2] Marti, K.: Approximationen der Entscheidungsprobleme mit linearer Ergebnisfunktion und positiv homogener, subadditiver Verlustfunktion, *Z. Wahrsch. verw. Geb.* **31** (1975), 203–233.

[3] Kall, P.: *Stochastic Linear Programming*, Springer-Verlag, Berlin, 1976.

[4] Kall, P. and Wallace, S. W.: *Stochastic Programming*, Wiley, Chichester, 1994.

[5] Mayer, J.: *Stochastic Linear Programming Algorithms*, Gordon and Breach, 1998.

[6] Marti, K.: Entscheidungsprobleme mit linearem Aktionen- und Ergebnisraum, *Z. Wahrsch. verw. Geb.* **23** (1972), 133–147.

[7] Marti, K.: Diskretisierung stochastischer Programme unter Berücksichtigung der Problemstruktur, *Z. Angew. Math. Mech.* **59** (1979), T105–T108.

[8] Marti, K.: Approximationen stochastischer Optimierungsprobleme, Verlag Anton Hain Meisenheim GmbH, Königstein/Ts., 1979.

[9] Marti, K.: Computation of descent directions in stochastic optimization problems with invariant distributions, *Z. Angew. Math. Mech.* **65** (1995), 355–378.

[10] Marti, K.: *Descent Directions and Efficient Solutions in Disretely Distributed Stochastic Programs*, Lecture Notes in Econom. Math. Systems 299, Springer-Verlag, Berlin, 1988.

[11] Marti, K.: Computation of efficient solutions of discretely distributed stochastic optimization problems, *Math. Methods Oper. Res.* **36** (1992), 259–294.

Chapter 12

THE STRUCTURE OF MULTIVARIATE MODELS AND THE RANGE OF DEFINITION

Vydūnas Šaltenis and Vytautas Tiešis
Institute of Mathematics and Informatics
Akademijos 4, 2600 Vilnius
Lithuania
saltenis;tiesis@ktl.mii.lt

Abstract The paper deals with a decomposition of a multivariate function into the summands of different dimensionality. The proposed methods of structure analysis enable to approximate the multidimensional function (the objective function in optimisation) by the functions of fewer variables. It is shown that step by step partition of the range of definition may be used to reduce the interactions of variables in the parts.

Keywords: Decomposition, structure analysis, approximation, optimisation

1. Introduction

Multidimensionality of complex models in global optimisation, classification, approximation or experiment design is one of the main obstacles to efficient solving of various practical problems.

It is attractive to simplify the multidimensional problems in order to reduce the dimensionality and to control the results of simplification. In order to simplify the analysis, and often as a matter of necessity, it is natural in a model not take certain variables explicitly into account or to hold them constant, even though they may be related to the phenomenon. Examples are easy to find.

Decomposition is a powerful tool for the analysis of large and complex systems. The technique of decomposing a system has been successfully used in many areas of engineering and science.

Simon and Ando [1] made two fundamental observations for dynamic systems:

1. Frequently complexity takes the form of a hierarchy, whereby a complex system is composed of interrelated subsystems that have in turn

G. Dzemyda et al. (eds.), Stochastic and Global Optimization, 209–219.
© 2002 *Kluwer Academic Publishers.*

their own subsystems, and so on, until some lowest level of elementary components is reached.

2. In general, interactions inside subsystems are stronger and/or more frequent that interactions among subsystems.

Courtois' [2,3] analysis is based on the works of Simon and Ando, and on the observations that large computing systems can be regarded as nearly completely decomposable systems.

There are many arguments to support these views.

We introduce a term "structure" for the possibilities to simplify the multidimensional problems in such a way. The problems of the structure are related to the possibility to approximate the multidimensional function (the objective function in optimisation) by the functions of fewer variables. For example, we can imagine that there exists a possibility to select some one-variable part of the function. This part, in a sense, approximates the multidimensional function after its simplification to the dimensionality one. In a similar way we can examine two-variable, three-variable, and s-dimensional parts ($s < n$).

The decomposition must meet some requirements:

(1) the uniqueness of decomposition;

(2) orthogonality of parts;

(3) optimality of decomposition.

The orthogonality is desirable because the influence between the parts is minimal in the case. The optimality requires that the parts of the decomposition would approximate the multidimensional function in the best way.

2. Decomposition into Components of Different Dimensionality

A decomposition of a multivariate function into the summands of different dimensionality [4–7] makes the basis for the structure analysis.

Let a function $f(X) = f(x_1, \ldots, x_n)$ be defined, for simplicity, on the cube $K^n(a_1 \leqslant x_1 \leqslant b_1, \ldots, a_n \leqslant x_n \leqslant b_n)$: $X \in K^n$. Sometimes a brief notation f will be used for $f(x_1, \ldots, x_n)$.

Let us introduce notation for the domain set of the function f which is a Cartesian product of basic domains $\Omega_1, \ldots, \Omega_n$: $\Omega = \Omega_1 \times \cdots \times \Omega_n$, and for special domains:

$$\Omega_{i_1 \ldots i_s} = \Omega_{i_1} \times \cdots \times \Omega_{i_s}, \quad 1 \leqslant i_1 < \ldots < i_s \leqslant n, \quad s = 1, \ldots, n,$$
$$\Omega_{(i)} = \Omega_1 \times \cdots \times \Omega_{i-1} \times \Omega_{i+1} \times \cdots \times \Omega_n, \quad i = 1, \ldots, n,$$
$$\Omega_{(ij)} = \Omega_1 \times \cdots \times \Omega_{i-1} \times \Omega_{i+1} \times \cdots \times \Omega_{j-1} \times \Omega_{j+1} \times \cdots \times \Omega_n,$$
$$i, j = 1, \ldots, n, i < j.$$

In the general case, the domain $\Omega_{(i_1 \dots i_s)}$ is defined in a similar way. The corresponding Lebesgue measures of domains are denoted by μ, $\mu_{i_1 \dots i_S}$ and $\mu_{(i_1 \dots i_S)}$.

Let f be a real-valued square integrable function: $f \in L_2(\Omega_1 \times \dots \times \Omega_n)$. The integral over the domain Ω is simply denoted as $\int_\Omega f$.

Let us denote the mean value of the function f by the same letter f with the upper index depending on the domain of integration. The mean value of f, when the integration is over all n basic domains except s basic domains: $\Omega_{i_1}, \dots, \Omega_{i_s}$, $1 \leqslant i_1 < \dots < i_s \leqslant n$, $s = 1, \dots, n$, is denoted as:

$$f^{i_1 \dots i_s} = \frac{1}{\mu_{(i_1 \dots i_s)}} \int_{\Omega_{(i_1 \dots i_s)}} f.$$

Naturally $f^{i_1 \dots i_s}$ is the s-variable function. For example, $f^i = \frac{1}{\mu_{(i)}} \int_{\Omega_{(i)}} f$ is the one-variable function.

We denote the mean value over the entire domain as

$$f^0 = \frac{1}{\mu} \int_\Omega f = \text{const}.$$

2.1. Functions Used in the Approximation

In the next sections we shall use some results and methods from Golomb [8].

We shall use the least-squares approximations, i.e. the approximations minimising the error:

$$\|f - f_e\| = \left(\int_\Omega (f - f_e)(f - f_e) \right)^{1/2},$$

where f_e is the approximating function.

In our case, f_e consists of the sum of functions of fewer variables. The dimensionality characterises the order of the approximating function. The approximating function of zero order $f_e^{(0)}$ is the constant. The first order approximating function

$$f_e^{(1)} = \sum_{i=1}^{n} f_i(x_i)$$

is the sum of one-variable functions. In the general case

$$f_e^{(s)} = \sum \cdots \sum_{1 \leqslant i_1 < \dots < i_s \leqslant n} f_{i_1 \dots i_s}(x_{i_1}, \dots, x_{i_s}), \quad s = 1, \dots, n-1,$$

where $f_{i_1 \dots i_s}(x_{i_1}, \dots, x_{i_s})$ are s-variable square integrable functions.

Each s-order approximation also contains approximations of lower orders. For example, $f_e^{(1)}$ contains $f_e^{(0)}$. It is more suitable to introduce "pure" approximating functions of s-order, which approximate the remainder

$$\psi_s = f - \sum_{i=0}^{s-1} f_e^{(i)}.$$

The next proposition is well known – the average is a least-square approximation:

$$f_e^{(0)} = \frac{1}{\mu}\int_\Omega f = f^0.$$

2.2. Approximation by Functions of a Single Variable

If an approximating function $f_e^{(1)} = \sum_{i=1}^{n} f_i(x_i)$, $f_i \in L_2(\Omega_i)$, is used to approximate the function $\psi_1 = f - f_e^{(0)}$, then the next proposition is true.

PROPOSITION 2.1. $f_i = \psi_1^i + c_i, i = 1, \ldots, n$, *where* c_i *are constants requiring that* $\sum_{i=1}^{n} c_i = 0$.

The proof is based on the idea that the first variation of the approximation error must be equal to zero. The proof of all propositions of this chapter in the case of unit cube K^n may be found in [5]. In the general case the proof needs some extra obvious simple algebraic calculations.

PROPOSITION 2.2. *The system of functions* $f_i, i = 1, \ldots, n$, *and* f^0 *is orthogonal if* $c_i = 0, i = 1, \ldots, n$.

2.3. The Approximation by Functions of Two Variables

The approximation is similar:

$$f_e^{(2)} = \sum_{i<j}^{n} f_{ij}(x_i, x_j), \quad i, j = 1, \ldots, n, \; f_{ij} \in L_2(\Omega_i \times \Omega_j),$$

is used to approximate the function $\psi_2 = f - f_e^{(1)} - f_e^{(0)}$, and the next proposition is true.

PROPOSITION 2.3. $f_{ij} = \psi_2^{ij} + c_{ij}$, *where* c_{ij} *are constants requiring that* $\sum_{i<j}^{n} c_{ij} = 0$.

The proof is similar to that of Proposition 2.1.

PROPOSITION 2.4. *The system of functions* f_{ij}, f_i, *and* f^0*is orthogonal if* $c_{ij} = 0, i, j = 1, \ldots, n, i < j$.

2.4. General Case of Approximation

If an approximating function

$$f_e^s = \sum_{1 \leqslant i_1 < \cdots < i_s \leqslant n} \cdots \sum f_{i_1 \dots i_s}(x_{i_1}, \dots, x_{i_s}),$$

$$s = 1, \dots, n, \; f_{i_1 \dots i_s} \in L_2(\Omega_{i_1} \times \cdots \times \Omega_{i_s}),$$

is used to approximate the function $\psi_s = f - \sum_{i=0}^{s-1} f_e^{(i)}$, then the next proposition is true.

PROPOSITION 2.5. *$f_{i_1 \dots i_s} = \psi_s^{i_1 \dots i_s} + c_{i_1 \dots i_s}$, $s = 1, \dots, n$, where $c_{i_1 \dots i_s}$ are constants requiring that*

$$\sum_{1 \leqslant i_1 < \cdots < i_s \leqslant n} \cdots \sum c_{i_1 \dots i_s} = 0.$$

PROPOSITION 2.6. *The system of functions $f_{i_1 \dots i_s}$, $s = 1, \dots, n$, and f^0 is orthogonal if $c_{i_1 \dots i_s} = 0$.*

2.5. The Structure Characteristics

The previous paragraphs give a possibility to decompose a multidimensional function into the summands of various dimensionality:

$$f = \sum_{i=0}^{n-i} f_e^{(i)} + f_{1 \dots n},$$

where $f_{1 \dots n}$ denotes an n-dimensional part, the remainder of the function f, which was not approximated by the functions of dimensionality lower than n.

We use groups of indices $i_1, \dots, i_s$, where $1 \leqslant i_1 < \cdots < i_s \leqslant n$, $s = 1, \dots, n$, and denote the sum with $2^n - 1$ terms as:

$$\sum^{\wedge} T_{i_1 \dots i_s} = \sum_{i=1}^{n} T_i + \sum_{1 \leqslant i < j \leqslant n} \sum T_{ij} + \cdots + T_{12 \dots n}.$$

Then the decomposition can be changed as follows:

$$f = f_0 + \sum^{\wedge} f_{i_1} \dots i_s(x_{i_1}, \dots, x_{i_s}). \tag{2.1}$$

The decomposition is unique and orthogonal for each function f integrable on K^n [6], if f_0 is constant and the integrals of summands (2.1) are equal to zero:

$$\int_{\Omega_{i_k}} f_{i_1} \dots i_s(x_{i_1}, \dots, x_{i_s}) \, \mathrm{d}x_{i_k} = 0, \quad 1 \leqslant k \leqslant s. \tag{2.2}$$

The summands of decomposition (2.1) may be found just like some integrals. Then, after integrating (2.1) on Ω, the constant summand will be equal to

$$f_0 = \frac{1}{\mu}\int_{\Omega} f. \tag{2.3}$$

One-dimensional summands, after integrating on $\Omega_{(i)}$, will be equal to

$$f_i(x_i) = f^i - f_0, \quad i = 1, \ldots, n, \tag{2.4}$$

two-dimensional summands, after integrating on $\Omega_{(ij)}$, will be equal to

$$f_{ij}(x_i, x_j) = f^{ij} - f_0 - f_i(x_i) - f_j(x_j), \quad i, j = 1, \ldots, n, i < j,$$

and so on.

Let us introduce the next characteristics of dispersion for the summands of the decomposition:

$$D_{i_1 \ldots i_s} = \frac{1}{\mu}\int_{\Omega} (f_{i_1} \ldots {}_{i_s})^2 = \frac{1}{\mu}\| f_{i_1} \ldots {}_{i_s}\|^2. \tag{2.5}$$

The decomposition is orthogonal, therefore,

$$D = \frac{1}{\mu}\hat{\sum}\| f_{i_1} \ldots {}_{i_s}(x_{i_1}, \ldots, x_{i_s})\|^2 = \hat{\sum} D_{i_1 \ldots i_s},$$

where

$$D = \frac{1}{\mu}\int_{\Omega} (f)^2 - (f_0)^2. \tag{2.6}$$

It is convenient to use the centred and normalised function f: $f^0 = 0$, $\|f\|^2 = \mu$. Then $\hat{\sum} D_{i_1 \ldots i_s} = 1, s = 1, \ldots, n$.

For normalised values $D_{i_1 \ldots i_s}$ we shall use the term *structure characteristics*. They are, in some sense, the measures of influence for separate variables or their groups. The first-order characteristics D_i, $i = 1, \ldots, n$, characterise the degree of influence for separate variables. The second-order characteristics D_{ij}, $i, j = 1, \ldots, n$, $i < j$, characterise the degree of influence for pairs of variables. Using the terms of variance analysis the second-order characteristics may be treated as *interactions*. Generally, an index s of structure characteristics $D_{i_1 \ldots i_s}$, $1 \leqslant i_1 < \cdots < i_s \leqslant n$, indicates the order of the characteristic.

Some part of the problems contains an additive error ξ, which is added to the values of the function. Usually, this error is the error of measurement of physical values, sometimes the error of computation. The situations with some

Table 1. Structure characteristics of the simple two-dimensional test functions.

$f(X)$	D_1	D_2	D_{12} (interaction)
$x_1 + x_2$	0.5	0.5	0
$x_1 x_2 - \dfrac{x_1 + x_2}{2}$	0	0	1

error arise, as a rule, in the problems of experiment design. In this case, the system of structural characteristics must be supplemented by an additional structure characteristic D_ξ of error. If the error is statistically independent, then the characteristic D_ξ is equal to the variance of the error: $D_\xi = D(\xi)$.

Table 1 contains the structure characteristics of the simple two-dimensional test functions, $\Omega_i = [0, 1]$.

2.6. Evaluation of the Characteristics

Usually we know only the values of function $f(X)$ for some points $X^j (j = 1, \dots, N)$. Then the Monte-Carlo method may be used for evaluations basing on (2.3), (2.5) and (2.6) if the co-ordinates of the points are uniformly distributed in Ω:

$$f_0 \approx \frac{1}{N} \sum_{j=1}^{N} f(X^j),$$

$$D + (f_0)^2 \approx \frac{1}{N} \sum_{j=1}^{N} \left(f(X^j)\right)^2.$$

s co-ordinates $Y^j = (x_{i_1}, \dots, x_{i_s})$ must be identical for pairs of random points used for the evaluation of the structure characteristics $D_{i_1 \dots i_s}$ [6]:

$$D_{i_1 \dots i_s} + (f_0)^2 \approx \frac{1}{N} \sum_{j=1}^{N} f(Y^j, Z^j) f(Y^j, U^j),$$

where Y^j are random points of dimensionality s, uniformly distributed in $\Omega_{i_1 \dots i_s}$, Z^j and U^j are random points of dimensionality n-s, uniformly distributed in $\Omega_{(i_1 \dots i_s)}$.

The structure analysis of some complex multidimensional models and applications of the structure characteristics are presented in [9–11].

3. An Interaction in an Infinitely Decreasing Range of Definition

In this section we will prove that it is always possible to select a sufficiently small range in a neighbourhood of a non-stationary point so that the ratio $\sum_{i=1}^{n} D_i/D$ of structure characteristics in the range would be arbitrarily near to 1. In other words, all variable interactions may be reduced as much as desired by reducing the range of definition. Let us describe the shrinking range K^n by constants r_i and the shrinking variable $\Delta = (b_i - a_i)/r_i$.

Let the function $f(X)$ be twice uniformly differentiable function in a closed range Ω. Then all second derivatives of the function $f(X)$ will be bounded. We will evaluate the structure characteristics under this condition. In order to calculate the integrals (2.3), (2.5) and (2.6), we will decompose the function $f(X)$ into Taylor series around the central point X_c, $x_{ci} = (a_i + b_i)/2$ of the range:

$$f(X) = f(X_c) + \sum_i \left.\frac{\partial f}{\partial x_i}\right|_{X_c} (x_i - x_{ci}) + \\ + \frac{1}{2}\sum_i \sum_j \left.\frac{\partial f}{\partial x_i \partial x_j}\right|_{\xi(X)} (x_i - x_{ci})(x_j - x_{cj}), \tag{3.1}$$

where $\xi(X) \in [X, X_c]$. Here and farther the indexes i, j of sums vary in the range: $i, j = 1, \ldots, n$. The integrals of remainder terms will be evaluated by the use of generalised mean value theorem replacing the values of derivatives by the bounded mean value.

PROPOSITION 3.1. *$f_0 = f(X_c) + R_0\Delta^2$, where $\Delta = (b_i - a_i)/r_i$, r_i are constants, R_0 is bounded for any bounded sequence of Δ.*

SKETCH OF PROOF. Let us calculate f_0:

$$f_0 = \frac{1}{\mu}\int_\Omega f(X) = f(X_c) + \sum_i \left.\frac{\partial f}{\partial x_i}\right|_{X_c} \frac{1}{b_i - a_i}\int_{a_i}^{b_i} (x_i - x_{ci})\,\mathrm{d}x_i + \\ + \frac{1}{2}\sum_i \frac{1}{b_i - a_i}\int_{a_i}^{b_i} \left[\frac{1}{\mu(i)}\int_{\Omega(i)} \left.\frac{\partial^2 f}{\partial^2 x_i}\right|_{\xi(X)}\right](x_i - x_{ci})^2\,\mathrm{d}x_i + \\ + \sum_i \frac{1}{b_i - a_i}\sum_j \frac{1}{b_j - a_j}\int_{a_i}^{b_i}\int_{a_j}^{b_j}\left[\frac{1}{\mu(ij)}\left[\int_{\Omega(ij)} \left.\frac{\partial^2 f}{\partial x_i \partial x_j}\right|_{\xi(X)}\right] \times \\ \times (x_i - x_{ci})(x_j - x_{cj})\right] \mathrm{d}x_i \mathrm{d}x_j. \tag{3.2}$$

Let us calculate the terms of formulae (3.2). The integral in the second term equals to zero. The generalised mean value theorem is used to calculate the integral of third term; the continuous derivative is changed by its mean value:

$$\eta_i = \left.\frac{\partial^2 f}{\partial^2 x_i}\right|_\nu, \quad \nu \in \Omega.$$

So the third term is equal to

$$\frac{1}{2}\sum_i (b_i - a_i)^2 \eta_i .$$

Concerning the fourth term the generalised mean value theorem is used for quarters of the region Ω_{ij} where the expressions $(x_i - x_{ci})(x_j - x_{cj})$ have uniform sign. Let us denote

$$\eta_{ij}^k = \left.\frac{\partial^2 f}{\partial x_i \partial x_j}\right|_v ,$$

where ij-th co-ordinates of the vector v depend to the k-th quarter of Ω_{ij} enumerated in usual counter clockwise order. Then the fourth term is equal to:

$$\frac{1}{64}\sum_i \sum_j (b_i - a_i)(b_j - a_j)(\eta_{ij}^1 + \eta_{ij}^3 - \eta_{ij}^2 - \eta_{ij}^4).$$

We get

$$f_0 = f(X_c) + R_0 \Delta^2, \tag{3.3}$$

after putting Δ and terms' expressions into (3.2). There R_0 is bounded and depends on continuous and bounded derivatives of second order. Proposition 3.1 is proved.

In order to evaluate the characteristic D the expressions (3.1) and (3.3) are put into (2.6). The integrals are evaluated similarly as in the previous proof and the next proposition may be formulated:

PROPOSITION 3.2.

$$D = \frac{\Delta^2}{12}\sum_i \left(\left.\frac{\partial f}{\partial x_i}\right|_{X_c}\right)^2 r_i^2 + o(\Delta^2). \tag{3.4}$$

The approximating function f_k is evaluated in the similar way:

$$\begin{aligned} f_k(x_k) = {} & \frac{1}{\mu_{(k)}} \int_{\Omega_{(k)}} (f(X) - f_0) = \left.\frac{\partial f}{\partial x_k}\right|_{X_c} (x_k - x_{ck}) + \\ & + 0.5 \left.\frac{\partial^2 f}{\partial^2 x_k}\right|_{\xi(x_k)} (x_k - x_{ck})^2 + (x_k - x_{ck}) R_1(x_k)\Delta + \\ & + R_2(x_k)\Delta^2 + R_0\Delta^2, \end{aligned} \tag{3.5}$$

where $R_1(x_k)$, $R_2(x_k)$ are continuous and bounded functions in an interval $[a_k, b_k]$, depending on second order derivatives of f; and the k-th co-ordinate of the mean value's argument $\xi(x_k) \in \Omega$ depends to $[x_k, x_{ck}]$.

In order to evaluate the characteristic D_k the expression (3.5) is integrated and the mean value theorem is used for the functions $R_1(x_k)$, $R_2(x_k)$ in the next proposition.

PROPOSITION 3.3. *The evaluation of first-order characteristics is*

$$D_k = \frac{1}{\mu}\int_{\Omega} [f_k(x_k)]^2 = \left(\frac{\partial f_k}{\partial x_k}\bigg|_{X_c}\right)^2 \Delta^2 r_k^2/12 + o(\Delta^2) \qquad (3.6)$$

The main result of the section is the next theorem. It is got by division (3.6) by (3.4).

THEOREM 3.1. *At a neighbourhood of a non-stationary point (where* $\sum_i (\frac{\partial f}{\partial x_i}|_{X_c}) \neq 0$) *twice-differentiable function tends to be separable that is to be represented by summands less than a second order:*

$$\lim_{\Delta\to 0} \sum_{i=1}^{n} D_i/D = 1.$$

PROOF. From Propositions 3.2 and 3.3 we have

$$\frac{D_k}{D} = \frac{\left(\frac{\partial f_k}{\partial x_k}\big|_{X_c}\right)^2 \Delta^2 r_k^2 + o(\Delta^2)}{\sum_i \left(\frac{\partial f_i}{\partial x_i}\big|_{X_c}\right)^2 \Delta^2 r_i^2 + o(\Delta^2)} = \frac{\left(\frac{\partial f_k}{\partial x_k}\big|_{X_c}\right)^2 r_k^2 + O(\Delta)}{\sum_i \left(\frac{\partial f_i}{\partial x_i}\big|_{X_c}\right)^2 r_i^2 + O(\Delta)}$$

$$\underset{\Delta\to 0}{\longrightarrow} \frac{\left(\frac{\partial f_k}{\partial x_k}\big|_{X_c}\right)^2 r_k^2}{\sum_i \left(\frac{\partial f_i}{\partial x_i}\big|_{X_c}\right)^2 r_i^2}.$$

Summing by k we have $\sum_k \frac{D_k}{D} \underset{\Delta\to 0}{\longrightarrow} 1$.

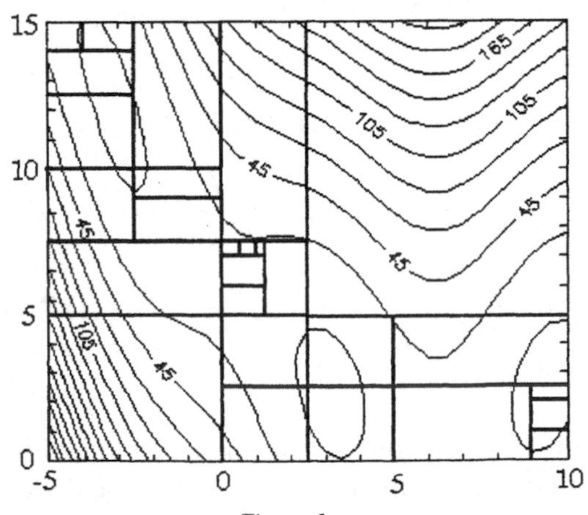

Figure 1.

4. The Example of Partition of the Range of Definition

We shall use a widely used in optimisation Branin's test function [12] of two variables. Figure 1 illustrates the level lines of this function (thin lines). It is obvious that the interaction of the variables of the function in whole range of definition is relatively strong (the second-order structure characteristic which characterises the interaction is equal to $D_{12} = 0.60$). Step by step partition of the range of definition was used to reduce the interactions in the parts to the level $D_{12} = 0.03$. The thicker lines in Figure 1 illustrate this possible partition.

5. Conclusions

The smaller the range of definition the smaller is an interaction between the variables. The fact may be used to find some sub-regions with small interactions, and to increase the efficiency of optimisation methods.

Bibliography

[1] Simon, H. A. and Ando, A.: Aggregation of variables in dynamic systems, *Econometrica* **29** (1964), 111–138.

[2] Courtois, P.-J.: *Decomposability*, Academic Press, New York, 1977.

[3] Courtois, P.-J.: On time and space decomposition of complex structures, *Comm. ACM* **28**(6) (1985), 590–603.

[4] Cukier, R. I., Levine, H. B. and Shuler, K. E.: Nonlinear sensitivity analysis of multiparameter model systems, *J. Comput. Phys.* **26**(1) (1978), 1–42.

[5] Šaltenis, V.: *Structure Analysis of Optimisation Problems*, Mokslas, Vilnius, 1989, 123 p. (in Russian).

[6] Sobol', I. M.: On sensitivity estimation for nonlinear mathematical models, *Mat. Mod.* **2**(1) (1990), 112–118 (in Russian).

[7] Sobol', I. M.: Global sensitivity indices for nonlinear mathematical models and their Monte Carlo estimates, *Math. Comput. Simulation* **55** (2001), 271–280.

[8] Golomb, M.: Approximation by functions of fewer variables, In: *On Numerical Approximation. Proceedings of a Symposium, Conducted by the Mathematics Research Centre*, R. E. Langer (ed.), The University of Wisconsin Press, Madison, 1959, pp. 275–327.

[9] Šaltenis, V.: Analysis of multivariate function structure in classification problems, *Informatica* **7**(4) (1996), 525–541.

[10] Šaltenis, V.: Grid with uniformity adapted to the structure of a multidimensional problem, *Informatica* **8**(4) (1997), 583–598.

[11] Šaltenis, V.: Global sensitivity analysis of infection spread, radar search and multiple criteria decision models, *Informatica* **9**(2) (1998), 235–252.

[12] Dixon, L. C. W. and Cziego, G. P.: *The Global Optimisation Problem: An Introduction*, Towards Global Optimisation 2, North-Holland, Amsterdam, 1978, pp. 1–15.

Chapter 13

OPTIMALITY CRITERIA FOR INVESTMENT PROJECTS UNDER UNCERTAINTY

Sergey A. Smolyak
CEMIRAS
Nakhimovsky prospekt 47
117418 Moscow, Russia
sergey36@yandex.ru

Abstract The uncertainty of effect of investment projects can have various types. Known types, namely, set-uncertainty and probabilistic one we consider as special cases of new type set-probabilistic uncertainty. Under such uncertainty the effect of the investment project is random variable with not exactly known distribution. We formalize such projects as families of one-dimensional probability distributions. Then the criterion for projects comparison becomes some functional on a class of distributions families. To ensure rational economic behavior of firm in which the decisions on projects selection are decentralized, such functional should be monotonous, continuous and additive. It turned out that such functional is generalization of mean criterion and Hurwicz's criterion (average weighted of extremal means of distributions included in family).

Keywords: Investment project, efficiency, evaluation, various types of uncertainty, comparison criteria, expected effect, axiomatics

During the elaboration of investment projects the tasks arising are project efficiency evaluating and choosing the best out of several alternative ones. In a deterministic situation the NPV index is usually used as the efficiency criterion. It reflects the discounted difference between the project revenues and expenditures. An investment project is considered effective in case its NPV (total discounted effect) is nonnegative. The greater is the project's NPV, the better. Such criterion has characteristics of monotonicity (it grows at increase of project's results or decrease of its expenses) and additivity (at joint realization of the independent projects their effects are added).

However usually projects are designed under uncertainty conditions, and the problem of its comparison becomes not trivial. The criterion should here reflect all possible values of NPV of the project. We shall call such index by *expected effect*. The expected effect of the project X we shall denote as $\mathbf{E}(X)$.

G. Dzemyda et al. (eds.), Stochastic and Global Optimization, 221–233.
© 2002 *Kluwer Academic Publishers.*

The problem is to discover how this criterion is arranged. Its structure depends on a type of uncertainty, which can be various. Some of such types we shall consider in the given paper. For solution of this problem we use the axiomatic approach which, however, differs from conventional in utility theory.

1. Probabilistic Uncertainty

At probabilistic uncertainty the project effect is random variable having known distribution. This situation was considered yet in [1] and its research has brought eventually to the theory of expected utility [2]. From this theory it follows that expected effect of the project, NPV of which has distribution P, looks like:

$$\int_{-\infty}^{\infty} u(x)\,\mathrm{d}P(x), \tag{1}$$

where $u(x)$ is utility function of determined effect x prescribed exogeneously. For $u(x) \equiv x$ or local projects such criterion is equivalent to the criterion of mathematical expectation of NPV widely used in design practice and financial management [3,4, and others]. However the initial information on utility functions is usually inaccessible for designers. Therefore neither criterion (1) nor more general criteria received in the subsequent papers [5–7, and others] are practically used.

The situation when the project effects depend on unknown "state of world" is considered in [8]. In other words, we have here "external" uncertainty that is identical to all projects. Here the criterion for comparison of the projects also looks like mathematical expectation of utility of effect, but now it is determined on a subjective probabilistic measure. Being based on it, some designers make the small (local) projects evaluation using subjective probabilities for taking account of their uncertainty (it is not a common rule).

We can frequently observe other situation, when each of estimated/compared projects is characterized by the own "internal" uncertainty (in our former papers [9,10] this situation was investigated differently). In conditions of external uncertainty the effects of different projects depend from each other, as they are determined by the same state of world. Under "internal" uncertainty the independent projects exist and their random effects are generated by different "mechanisms" and are independent.

Distribution functions (DF) on real axis we shall designate P, Q or R. We denote

$$E_Q\big[u(x)\big] = \int_{-\infty}^{\infty} u(x)\,\mathrm{d}Q(x); \qquad E_Q = \int_{-\infty}^{\infty} x\,\mathrm{d}Q(x).$$

The distance between DFs Q and R we define by the formulae:

$$\rho(Q,R) = \big|E_Q - E_R\big| + \int_{-\infty}^{+\infty} |x|^r \big|Q(x+E_Q) - R(x+E_R)\big|\,\mathrm{d}x, \quad 1 \leqslant r < 2.$$

The function of degenerate distribution concentrated in the point s we denote by B_s and family including only one distribution B_s we denote by $\mathbf{B}_s$. Such family corresponds to the project with determined NPV $= s$. We call DF ρ*-bounded* if $\rho(Q, B_0) < \infty$.

Let us say that the distribution Q *dominates* distribution $R(Q \gg R)$ if the first stochastic domination condition is fulfilled: $Q(x) \leqslant R(x)\ \forall x$. If $Q \gg R$, $u(x)$ is a bounded increasing function, then $E_Q[u(x)] \geqslant E_R[u(x)]$.

Let **F** is class of ρ-bounded DF on real axis, on which are defined the domination relation ($\gg$), the convolution operation ($\oplus$) and distance ρ. Let's consider the projects which effects have distributions from **F**. The criterion of expected effect of such projects becomes a functional $\mathbf{E}(Q)$ on **F**. The following of its properties are deduced from economic reasons: (1) the expected effect of determined project (degenerated distributuion) coincides with its "usual" effect (NPV): $\mathbf{E}(B_s) = s$; (2) monotonicity: $P \gg Q \Rightarrow \mathbf{E}(P) \geqslant \mathbf{E}(Q)$; (3) at joint realization of independent projects their expected effects are summarized: $\mathbf{E}(P \oplus Q) = \mathbf{E}(P) + \mathbf{E}(Q)$.

However for our purposes it is not enough and, using idea [11] we shall enter the stronger requirement. A function $\omega(t)$ is called *good* if $\omega(t) \to 0$ at $t \to 0$, $t^{-2/r}\omega(t) \to 0$ at $t \to \infty$. Now we can describe the *uniform continuity* of criterion by the following *axiom*: there exists a good function $\omega(t)$ such that $|\mathbf{E}(X) - \mathbf{E}(Y)| < \omega(\rho(X, Y))$.

THEOREM 1. *Only functional* $\mathbf{E}(P) = E_P$ *has the indicated properties.*

This result confirms legitimacy of use of the traditional optimality criterion of projects with random effect. Theorem 1 is a corollary of more common statement which will be given in Sections 4–6.

2. Set-uncertainty

We can frequently observe other type of internal uncertainty. Here we know set X of possible values of NPV of project X, but we do not know probability distribution on this set. Such uncertainty we call *set-uncertainty*. Let **S** denote the class of bounded subsets of real axis. Now a project X is completely characterized here by corresponding set $X \in \mathbf{S}$. Let's introduce the domination relation on **S**: $X \gg Y$, if any $x \in X$ is no less than some $y \in Y$, and any $y \in Y$ is no more than some $x \in X$. The joint realization of projects can be formalized as the Minkowsky-summing of sets: $X \oplus Y = \{x + y \mid x \in X, y \in Y\}$. The expected effect criterion can be considered as a functional $\mathbf{E}(X)$ on **S**.

This functional also has properties (1)–(3): (1) for the determined projects (single-point X) the expected effect coincides with "usual" effect: $\mathbf{E}(\{b\}) = b$; (2) monotonicity: $X \gg Y \Rightarrow \mathbf{E}(X) \geqslant \mathbf{E}(Y)$; (3) at joint realization of the independent projects their expected NPVs are summarized: $\mathbf{E}(X \oplus Y) = \mathbf{E}(X) + \mathbf{E}(Y)$.

The following result can be easily proved (its n-dimensional extension is given in [12]).

THEOREM 2. *Any functional on* **S** *possessing indicated properties* (1)–(3) *is L. Hurwicz's* [13] *criterion:*

$$\mathbf{E}(X) = \lambda \sup_{x \in X} x + (1 - \lambda) \inf_{x \in X} x \quad (0 \leqslant \lambda \leqslant 1). \tag{2}$$

Such criterion is used in design practice and even is reflected in the Russian normative documents on investment projects evaluation. However, criteria of a more general kind offered in [14] for the given situation, are not practically used.

3. Why the Considered Models of Uncertainty Are Unsuitable for Design Practice?

Why the design practice rejects the specified methods of taking uncertainty into account? We can give the following reasons:

- the offered formulae include subjective probabilities, utility functions or other information which is hard (for designers) to receive and also to prove and protect before independent experts, creditors and state bodies;
- the offered criteria are oriented at most to evaluate of the whole investment policy of firm not to decide local questions (whether it is favorable to build the given object? which variant of replacement of the worn out equipment is better? and so forth). They are poorly adapted to evaluate the project irrespectively to whole other activity of firm that realizes it. This creates significant difficulties for the designers, especially, if the information about "other" activity of firm makes a trade secret. Axiomatics conducting to such difficulties, in our opinion, is inadequate;
- the designers understand that many of project parameters are random variables. However, their probabilistic distributions are precisely unknown and the maximal value of each parameter is strongly differ from the minimal one;
- the project developers deal with such uncertainty that is not stacked in the "Procrustean bed" of offered types.

In the given work we consider other type of uncertainty, we show other requirements to criteria of expected effect and we find out structure of the appropriate criteria.

Thus, considering the projects with random NPV we assume that incomes/outcomes arising at different times are reduced to the present time with use of *risk-free* discount rate. In other words, we want to take into account risk of project by choosing a suitable method of aggregation of possible values of NPV but not by updating these values (in the financial analysis the randomness of incomes/outcomes usually taken into account by addition of risk premium to discount rate: see [4] for example; however, such method is inexact and will not be considered).

4. Set-probabilistic Uncertainty. Definitions

Really, the project NPV depends on dynamics of the market prices and demand for production and can frequently be considered as random variable. However, to estimate such project it is required not only to recognize the randomness of NPV but also to specify its distribution. Here the designer collides with impossibility to receive the necessary information for an establishment and a justification of the kind of distribution. Usually they have only retrospective data and several variants of the forecast. Similar situation we can describe by the following model of *set-probabilistic uncertainty*.

NPV of the project X in this model is characterized by some probability distribution, the information about which does not allow to establish it uniquely. It means, that NPV of the project X is characterized by not one DF but some family $X = \{Q\}$ of possible (allowable, coordinated with the initial information) distributions. Thus, real project we identify to family of DF on real axis and treat its expected effect as some functional **E** on the class of such families. We shall note two important special cases:

- if the family X consists of only one DF, the uncertainty turns in "usual", probabilistic one;

- if the family X consists of all DF on some bounded set H we have the Hurwicz's situation set-uncertainty: we know only set H of possible values of NPV and do not know anything of probability distribution on this set.

Further we shall restrict the class of distributions' families and establish the properties of functional **E** on this class ensuring rational economic behavior of the investor. It will allow to reveal structure of the specified functional. We shall enter the following definitions and designations.

The distance between projects X and Y we shall define in the sense of Hausdorff:

$$\rho(X, Y) = \max\left\{\sup_{P\in X}\inf_{Q\in Y}\rho(P, Q), \sup_{P\in Y}\inf_{P\in Q}\right\}.$$

We call a family X *ρ-bounded* if there is constant C such that $\rho(Q, B_0) < C$ for all DF $Q \in X$. Further under the projects we shall understand only *ρ-bounded* families of DFs.

Let us say that the project X *dominates* project Y ($X \gg Y$), if:

- each DF $P \in X$ dominates at least one DF $Q \in Y$;
- each DF $Q \in Y$ is dominated by at least one DF $P \in X$.

If the family X includes only one DF P and Y includes only one DF Q the conditions $X \gg Y$ and $P \gg Q$ are equivalent.

The projects $X_1, \ldots, X_m$ we call *independent* if any possible joint distribution of their NPV looks like $\prod_{k=1}^{m} P_k(x_k)$, where $P_k(x_k)$ is possible DF of NPV of X_k. Further, considering a few projects, we shall by default mean their *independence*.

We can represent project X as family of random variables with the same DF. However, two families of random variables $X = \{\xi\}$ and $Y = \{\eta\}$ with the same DF must be interpreted as different but equieffective projects. In this case we can call one of these sets the *copy* of the other. Later in this paper we will use the same letter to designate a project and its copies.

The basic operation on families is the summation operation that formalizes the result of joint realization of the independent projects. As already was marked, at joint realization of deterministic independent projects their NPVs are added. At summation of independent random variables the distribution of a sum is the convolution ($\otimes$) of distributions composed variables. Therefore, the sum of families X and Y is defined by $X + Y = \{R \mid R{=}P \otimes Q, P \in X, Q \in Y\}$.

In language of random variables it means that possible effects of project Z are sums of independent possible random effects of projects X and Y. Easily to check that the sum of *ρ-bounded* families will be *ρ-bounded* one, too. We shall note also, that at convolution of DFs their means and dispersions are summarized.

5. Axiomatics

To ensure economically rational behavior of the investor under set-probabilistic uncertainty such expected effect functional can only be that which satisfies to the proper requirements (axioms).

1. If uncertainty is absent and project NPV is precisely equal to s, it is natural to assume that the expected effect of such project is equal to s, too. It is expressed by the following *coordination axiom*: $\mathbf{E}(B_s){=}s$, $\forall s$.
2. We shall consider two projects having random NPV with DFs P and Q ($P \gg Q$). Then at any M the inequality NPV $> M$ is more probable

for the first project than for the second one. It is natural to assume that the first project thus will be not less effective than second one. In conditions of set-probabilistic uncertainty this requirement can be transformed as following *monotonicity axiom*: $X \gg Y \Rightarrow \mathbf{E}(X) \geqslant \mathbf{E}(Y)$.

3. The expected effect of project should vary insignificantly at small changes of possible distributions of NPV. Therefore, if the sequence of the projects X_n converges to X in metric ρ, then the expected effects $\mathbf{E}(X_n)$ should converge to $\mathbf{E}(X)$. Such requirement appears too weak, and we shall assume the *uniform continuity* of the functional: $|\mathbf{E}(Q) - \mathbf{E}(R)| < \omega(\rho(Q, R))$ at some good function ω.
4. The last axiom is necessary to ensure an opportunity of decentralized decision-making concerning a choice of the most effective projects. In the utility theory (see, for instance, [5]) it is called by *independence axiom*. In the beginning we shall give its pithy formulation suitable for any types of uncertainty and even in a determined case:

Let one project is more effective than the second, and the third project is independent from first two. Then the joint realization of the first and the third projects is not less effective, than joint realization of the second and the third ones.

$$\mathbf{E}(X) > \mathbf{E}(Y) \Rightarrow \mathbf{E}(X + Z) \geqslant \mathbf{E}(Y + Z), \quad \forall Z. \tag{3}$$

Usually, conditions of such kind are called *independence axioms* in the utility theory. The expected utility criteria of type (1) at $u(x) \neq cx$ do not satisfy just to this condition.

This axiom reflects the possibility of carrying out a local comparison of projects in principle. Following our works [12,15], we have chosen the above denomination in order to emphasize a possibility of the following way of justification.

Suppose (3) fails, that is, there exist X, Y, Z such that $\mathbf{E}(X) > \mathbf{E}(Y)$ but $\mathbf{E}(X + Z) < \mathbf{E}(Y + Z)$. Imagine now a company that has two branches, and the first branch selects for realization one of the projects X and Y. Let project Z be chosen by the second branch. Clearly, the manager of the first branch would select project X whereas the CEO of the whole company would prefer the "combination" of Y and Z to that of X and Z. So, the comparison of projects is not possible on lower or local levels, and the right of decision-making should be passed to the top level of management. In other words, the violation of (3) is incompatible with a decentralization of management in complicated production processes or economic systems.

Observe now that, when monotonicity and coordination axioms accepted, the condition (3) may be replaced by a simpler and more obvious one. It follows from (3) that $\mathbf{E}(X) = \mathbf{E}(Y) \Rightarrow \mathbf{E}(X + Z) = \mathbf{E}(Y + Z)$. Let $x = \mathbf{E}(X)$,

$z = \mathbf{E}(Z)$. Then we have $x = \mathbf{E}(B_x) = \mathbf{E}(X)$, $z = \mathbf{E}(B_z) = \mathbf{E}(Z)$. It follows that $\mathbf{E}(X + Z) = \mathbf{E}(B_x + Z) = \mathbf{E}(B_x + B_z) = \mathbf{E}(B_{x+z}) = x + z$. Thus the "reasonable" expected effect functionals are *additive*:

$$\mathbf{E}(X + Z) = \mathbf{E}(X) + \mathbf{E}(Z). \tag{4}$$

6. Main Theorem and Examples

The structure of functionals satisfying the above-mentioned axioms is presented by the main theorem.

THEOREM 3. $\mathbf{E}(X)$ *satisfy the above mentioned axioms if and only if it has a form:*

$$\mathbf{E}(X) = \lambda \sup_{P \in X} E_p + (1 - \lambda) \inf_{P \in X} E_P, \quad \textit{where } 0 \leqslant \lambda \leqslant 1. \tag{5}$$

PROOF. See Appendix.

We see that the received expected effect criterion is a generalization of both mathematical expectation criterion and Hurwicz's criterion (2). This criterion is recommended by the Russian Government for use at investment projects evaluation [16].

We give three examples of use of criterion (5). In these examples as well as in design practice, only finite number of the possible scenarios of the project realization is considered. Therefore the appropriate probability distributions are discrete and can be described by finite-dimensional vectors of probabilities.

EXAMPLE 1. NPV of project X can have 4 possible values: 100, 400, −60 and −240. It is known that the first of them is the most probable and the latter is the least probable. At known probabilities p_i of these values the expected NPV of the project is set by the formula

$$100p_1 + 400p_2 - 60p_3 - 240p_4. \tag{6}$$

However the initial information on probabilities p_i is expressed by only following formulas:

$$p_1 \geqslant p_2 \geqslant p_4 \geqslant 0; \qquad p_1 \geqslant p_3 \geqslant p_4; \qquad p_1 + p_2 + p_3 + p_4 = 1. \tag{7}$$

It determines the appropriate family X of probability distributions (p_1, p_2, p_3, p_4). Easily to be convinced that the expression (6) under conditions (7) has maximum equal to 250 ($p_1 = p_2 = 0.5$; $p_3 = p_4 = 0$) and minimum equal to 50 ($p_1 = p_2 = p_3 = p_4 = 0.25$). Therefore at $\lambda = 0.3$ we have $\mathbf{E}(X) = 0.3 \times 250 + 0.7 \times 50 = 110$.

EXAMPLE 2. Let n scenarios of the project X realization are possible and NPV at i-th scenario is $v_i (v_1 < v_2 < \cdots < v_n)$. Let it is known that extreme values of NPV are less probable than all other: $p_1, p_n \leqslant p_2, \ldots, p_{n-1}$. These inequalities determine the family X of probability distributions and the appropriate set of means of NPV namely $\sum_i v_i p_i$. Solving simple linear programming problems we find extreme values of these means and, by virtue of (5), we receive:

$$\mathbf{E}(X) = \lambda \max\left\{ \nu_{n-1}, \frac{\nu_2 + \cdots + \nu_n}{n-1} \right\} + (1-\lambda) \min\left\{ \nu_2, \frac{\nu_1 + \cdots + \nu_{n-1}}{n-1} \right\}.$$

EXAMPLE 3. At initial stage of the project X development the price of made new product is not exactly known and duration of construction is the same. Therefore designers have considered I possible values of the product price and J possible values of duration of construction. They have found out that in the scenario S_{ij} (at i-th value of the product price and j-th value of duration of construction) the NPV of the project is equal to v_{ij}. The probabilities q_{ij} of scenarios are unknown but on data of construction of similar objects it is established that j-th value of duration of construction has probability p_j.

Here the family X includes all joint distributions of product price and duration of construction, i.e., all IJ-dimensional vectors with components q_{ij} subject to $q_{ij} \geqslant 0$; $\sum_i q_{ij} = p_i (i = 1, \ldots, J)$. Extreme values of mathematical expectation of NPV ($\sum_{i,j} v_{ij} q_{ij}$) at these resrictions are easily calculated. In result we find:

$$\mathbf{E}(X) = \sum_j p_j \Big[\lambda \max_i \nu_{ij} + (1-\lambda) \min_i \nu_{ij} \Big]. \tag{8}$$

Let now we receive additional information that random fluctuations of product price and duration of construction are *independent*. Now the probabilities of scenarios become *interconnected*: $q_{ij} = \pi_i p_j$, where π_i is (unknown) probability of i-th value of product price. Here the value of expected effect will be essentially different from (8):

$$\begin{aligned} \mathbf{E}(X) &= \lambda \max_{\{\pi_i\}} \sum_{i,j} \pi_i p_j \nu_{ij} + (1-\lambda) \min_{\{\pi_i\}} \sum_{i,j} \pi_i p_j \nu_{ij} \\ &= \lambda \max_i \sum_j p_j \nu_{ij} + (1-\lambda) \min_i \sum_j p_j \nu_{ij}. \end{aligned}$$

We see that practical calculations of project efficiency under set-probabilistic uncertainty with criterion (5) are possible, and they require minimal information about probabilities of various events and about independence of some events.

Appendix. Proof of the Main Theorem

The *trace* of distribution Q, denoted by $\operatorname{tr} X$, is defined to be a point on a plane with coordinates (E_Q, D_Q). The trace of project X, denoted by $\operatorname{tr} X$, is defined to be the set of traces of all distributions from X. Let $\operatorname{ctr} X$ is the convex hull of $\operatorname{tr} X$.

Let T_Q denote the third central absolute moment of the DF Q. A family X is called *strongly bounded* if

$$\gamma(X) = \inf\{C \mid |E_Q| \leqslant C,\ 1/C \leqslant D_Q \leqslant C,\ T_Q \leqslant C,\ \forall Q \in X\} < \infty.$$

LEMMA 1. *If* $\operatorname{tr} X = \operatorname{tr} Y$, *then* $\mathbf{E}(X) = \mathbf{E}(Y)$.

PROOF. It is easily seen that any ρ-bounded distribution is a limit (in the ρ-metric) of strongly bounded ones having the same trace. Therefore, we only need to prove lemma for strongly bounded families X and Y.

Suppose the lemma is false and $\gamma(X) < C$, $\gamma(Y) < C$, $\operatorname{ctr} X = \operatorname{ctr} Y$, but $|\mathbf{E}(X) - \mathbf{E}(Y)| = C_0 > 0$. Let us consider two projects formed by the summation of n copies of the projects X and Y: $X' = X + X + \cdots + X$; $Y' = Y + Y + \cdots + Y$. By additivity,

$$|\mathbf{E}(X') - \mathbf{E}(Y')| = n|\mathbf{E}(X) - \mathbf{E}(Y)| = nC_0. \tag{9}$$

We shall prove that such equality is impossible.

Let $P \in X'$, $H = (E_P/n, D_P/n)$. Hence $H \in \operatorname{ctr} X = \operatorname{ctr} Y$. From this it followsthat there is a triangle containing I, three vertexes of which (m_i, D_i), $i = 1, 2, 3$, belong to $\operatorname{tr} Y$. This gives:

$$|m_i| \leqslant C, C^{-1} \leqslant D_i \leqslant C, \quad \left(\frac{E_P}{n}, \frac{D_P}{n}\right) = \sum_{i=1}^{3} t_i(m_i, D_i),$$

$$t_i \geqslant 0, \ \sum_{i=1}^{3} t_i = 1$$

Let Q_i denote DFs from Y such that $\operatorname{tr} Q_i = (m_i, D_i)$. Then there are the integers n_i such that $\sum_{i=1}^{3} n_1 = n$, $|nt_i - n_i| < 1$. Define

$$Q = \underbrace{Q_1 \otimes \cdots \otimes Q_1}_{n_1 \text{ times}} \otimes \underbrace{Q_2 \otimes \cdots \otimes Q_2}_{n_2 \text{ times}} \otimes \underbrace{Q_3 \otimes \cdots \otimes Q_3}_{n_3 \text{ times}}.$$

By the above, we have:

$$(E_Q, D_Q) = \sum_{i=1}^{3} n_i(m_i, D_i), \qquad \| E_Q - E_P| < 3C, \qquad |D_Q - D_P| < 3C,$$

$$D_P > \frac{n}{C}, \ D_Q > \frac{n}{C}. \tag{10}$$

Let R and S denote normal DF such that tr $R = (E_P, D_P)$, tr $S = (E_Q, D_Q)$. Then we have: $\rho(P, Y) \leqslant \rho(P, Q) \leqslant \rho(P, R) + \rho(R, S) + \rho(S, Q)$. It is easily seen that $\rho(R, S) < C_1$ by (10). Let us now examine $\rho(S, Q)$.

We know that Q is DF of a sum of n independent random variables with distributions Q_i. The DF S is normal and has the same mean E_Q and dispersion D_Q. Let us apply the following result of [17]:

$$\left|Q(x+E_P)-S(x+E_P)\right| \leqslant \frac{A\sum_i T_i}{(D_Q)^{3/2}\left(1+|x/\sqrt{D_Q}|^3\right)} \leqslant \frac{AC^{5/2}}{\sqrt{n}\left(1+|x\sqrt{C/n}|^3\right)},$$

where T_i is the third absolute moment of Q_i, $T_i < C$, A is a constant. It follows easily that $\rho(S, Q) < C_2 n^{r/2}$. The inequality $\rho(P, R) < C_2 n^{r/2}$ is proved similarly.

From this inequalities we obtain $\rho(P, Y) < C_3 n^{r/2}$ for all $P \in Y'$. Similarly, $\rho(Q, X) < C_3 n^{r/2}$ for all $Q \in X'$. Therefore $\rho(X, Y) < 2C_3 n^{r/2}$ and, by continuity axiom, $|\mathrm{E}(X') - \mathrm{E}(Y')| < \omega[\rho(X, Y)] < \omega(2C_3 n^{r/2}) = o(n)$, contrary to (9). □

From Lemma 1 we conclude that the expected effect of project X depends only on convex set ctr X. It allows to "translate" the problem of project evaluation to the "language" of planar convex sets.

Let H be a bounded planar convex set and X be a project such that tr $X = H$. The expected effect of set H is defined by $\mathbf{E}(H) = \mathbf{E}(X)$. By virtue of a Lemma 1, $\mathbf{E}(H)$ does not depend on chosen X, so the offered definition is correct. It is easily seen that the constructed functional has the following properties.

(1) At summation of projects their traces are summarized in sense on Minkowski. Since $\mathbf{E}(X)$ is additive functional, it follows that $\mathbf{E}(H)$ is the same: $\mathrm{E}(H_1) + \mathbf{E}(H_2) = \mathbf{E}(H_1 + H_2)$.

(2) The set consisting only one point $(m, 0)$ corresponds to degenerate distribution B_m. Hence, by virtue of coordination axiom, we have $\mathbf{E}(\{(m, 0)\}) = m$.

(3) Let the distance between planar sets be defined in the Haussdorff's sense. Then the functional $\mathbf{E}$ is continuous in this metric. From this, in particular, it follows that any set H is equieffective with own closure.

We say that a point $U_1 = (m_1, D_1)$ *dominates* a point $U_2 = (m_2, D_2)$ and denote this by $U_1 \gg U_2$, if there are DFs P_i $(i = 1, 2)$ such that tr $P_i = U_i$ and $P_1 \gg P_2$. We say that a set F *dominates* a set H $(F \gg H)$, if each point of F dominates some point of H and each point of H is dominated by some point of F.

LEMMA 2. $F \gg H \Rightarrow \mathbf{E}(F) \geqslant \mathbf{E}(H)$.

PROOF. Let $U(m, D) \in F$. Then there is some point $U'(m', D') \in H$ dominated by U. It follows that there are DFs P_U and Q_U such that $\operatorname{tr} P_U = U$, $\operatorname{tr} Q_U = U'$ and $P_U \gg Q_U$. Similarly, for each point $V(m', D') \in H$ there are DFs R_V and S_V such that $\operatorname{tr} R_V \in F$, $\operatorname{tr} S_V = V$ and $S_V \gg R_V$. Let X denote the union of all pairs (P_U, S_V) and Y denote the union of all pairs (Q_U, R_V). Then we have $\operatorname{tr} X = F$, $\operatorname{tr} Y = H$ and $X \gg Y$. It follows that $\mathbf{E}(F) = \mathbf{E}(X) \geqslant \mathbf{E}(Y) = \mathbf{E}(H)$. □

Now the structure of entered domination relation is easily established.

Let us denote by $P_{a,m,D}$ the distribution that has atoms $D/(D + a^2)$ and $a^2/(D + a^2)$ in points $m - a$ and $m + D/a$. Then $\operatorname{tr} P_{a,m,D} = (m, D)$. It is easy to check that at $m > m'$ there exists positive numbers a and a' such that $P_{a,m,D} \gg P_{a',m',D'}$. Therefore from $m > m'$ it follows that $(m, D) \gg (m', D')$. Define

$$\alpha(F) = \sup_{(m,D)\in F} m, \qquad \zeta(F) = \inf_{m,D\in F} m, \tag{11}$$

where F is the bounded convex planar set. Let F be *open* set and H be *open* segment connecting the points $U = (\alpha(F), 1)$ and $V = (\zeta(F), 1)$ but not including them.

Then, by this definition and Lemma 1, each point from F dominates any point from H, and each point from H is dominated by some point from F, so $H \gg F$. Similarly, we receive $F \gg H$. From this, by Lemma 2, it follows that $\mathbf{E}(F) = \mathbf{E}(H)$. Therefore $\mathbf{E}(F)$ depends only on $\alpha(F)$ and $\zeta(F)$: $\mathbf{E}(F) = \varphi(\alpha(F), \zeta(F))$. By continuity, this equality is valid for arbitrary bounded sets too.

By the above, the function φ is continuous and is monotone on both variables. Besides, for one-point set $F = \{(m, 0)\}$ we have $\varphi(m, m) = \mathbf{E}(\{(m, 0)\}) = m$. It should be noted, at last, that at Minkowski-summation of sets the corresponding values $\alpha(F)$ and $\omega(F)$ are also summarized. Therefore the function φ is additive. It is easily seen that these properties have only the functions

$$\mathbf{E}(F) = \varphi(\alpha(F), \zeta(F)) = \lambda\alpha(F) + (1 - \lambda)\zeta(F), \quad 0 \leqslant \lambda \leqslant 1. \tag{12}$$

Substituting (11) and $F = \tilde{\operatorname{n}}\operatorname{tr} X$ into (12) we obtain the required formula (5). The theorem is proved. □

Acknowledgements

The author wishes to express his gratitude to V. Arkin, A. Slastnikov, V. Livshits and V. Rotar' for their active interest in the investigations and for many stimulating conversations. This work was supported by Russian Humanitarian Science Fund (project #01-02-00430) and Russian Foundation for Basic Researches (project #99-06-80138).

Bibliography

[1] Bernoulli, D.: *Specimen theoriae novae de mensura sortis*, Commentarii academiae scientarium imperialis Petropolitanae, 1730, 1731, 1738.

[2] von Neumann, J. and Morgenstern, O.: *Theory of Games and Economic Behavior*, Princeton University Press, 1947.

[3] Brigham, E. F. and Gapensky. L. C.: *Intermediate Financial Management*, Dryden Press, 1985.

[4] Brealey, R. A. and Mayers, S. C.: *Principles of Corporate Finance*, International edition, McGraw-Hill Inc., 1991.

[5] Fishburn, P. C.: *Utility Theory for Decision Making*, Wiley, New York, 1970.

[6] Fishburn, P. C.: *The Foundations of Expected Utility*, D. Reidel Publishing Company, Dordrecht, 1982.

[7] Fishburn, P. C.: Retrospective on the utility theory of von Neumann and Morgenstern, *J. Risk and Uncertainty* **2** (1989).

[8] Savage, L. J.: *The Foundation of Statistics*, Wiley, New York, 1954.

[9] Smolyak, S. A.: On comparison rules of some variants of economic actions under uncertainty, In: *Studies in Stochastic Control Theory and Mathematical Economics*, CEMI Academy of Sciences of the USSR, Moscow, 1981 (in Russian).

[10] Arkin, V. I. and Smolyak, S. A.: On the structure of optimality criteria in stochastic optimization models, In: V. I. Arkin, A. Shiraev and R. Wets (eds), *Stochastic Optimization Proceedings of the International Conference, Kiev, 1984*, Lectures Notes in Control and Inform. Sci., 1986.

[11] Rotar', V. I. and Sholomitsky, A. G.: On the Pollatsek–Tversky theorem on risk, *J. Math. Psychol.* **38**(3) (1994).

[12] Smolyak, S. A.: On the rules of comparison of alternatives with uncertain results, In: *Probabilistic Models of Mathematical Economics*, CEMI Academy of Sciences of the USSR, Moscow, 1990 (in Russian).

[13] Hurwicz, L.: Optimality criteria for decision making under ignorance, *Cowles Commission Papers* No. 270 (1951).

[14] Arrow, K. J. and Hurwicz, L.: An optimality criterion for decision–making under ignorance, In: *Uncertainty and Expectation in Economics*, Basil Blackwell and Mott, Oxford, 1972.

[15] Smolyak, S. A.: Comparison criteria for fuzzy alternatives, *Math. Social Sci.* **27** (1994), 185–202.

[16] *The Methodical Recommendations for Evaluation of Investment Projects Efficiency*, 2nd edn, The official publication, Economika, Moscow, 2000, 421 pp. (in Russian).

[17] Bikelis, A.: The estimations of remainder term in the central limit theorem, *Liet. Mat. Rink.* **6**(3) (1966), 323–346 (in Russian).

Nonconvex Optimization and Its Applications

1. D.-Z. Du and J. Sun (eds.): *Advances in Optimization and Approximation*. 1994 ISBN 0-7923-2785-3
2. R. Horst and P.M. Pardalos (eds.): *Handbook of Global Optimization*. 1995 ISBN 0-7923-3120-6
3. R. Horst, P.M. Pardalos and N.V. Thoai: *Introduction to Global Optimization* 1995 ISBN 0-7923-3556-2; Pb 0-7923-3557-0
4. D.-Z. Du and P.M. Pardalos (eds.): *Minimax and Applications*. 1995 ISBN 0-7923-3615-1
5. P.M. Pardalos, Y. Siskos and C. Zopounidis (eds.): *Advances in Multicriteria Analysis*. 1995 ISBN 0-7923-3671-2
6. J.D. Pintér: *Global Optimization in Action*. Continuous and Lipschitz Optimization: Algorithms, Implementations and Applications. 1996 ISBN 0-7923-3757-3
7. C.A. Floudas and P.M. Pardalos (eds.): *State of the Art in Global Optimization*. Computational Methods and Applications. 1996 ISBN 0-7923-3838-3
8. J.L. Higle and S. Sen: *Stochastic Decomposition*. A Statistical Method for Large Scale Stochastic Linear Programming. 1996 ISBN 0-7923-3840-5
9. I.E. Grossmann (ed.): *Global Optimization in Engineering Design*. 1996 ISBN 0-7923-3881-2
10. V.F. Dem'yanov, G.E. Stavroulakis, L.N. Polyakova and P.D. Panagiotopoulos: *Quasidifferentiability and Nonsmooth Modelling in Mechanics, Engineering and Economics*. 1996 ISBN 0-7923-4093-0
11. B. Mirkin: *Mathematical Classification and Clustering*. 1996 ISBN 0-7923-4159-7
12. B. Roy: *Multicriteria Methodology for Decision Aiding*. 1996 ISBN 0-7923-4166-X
13. R.B. Kearfott: *Rigorous Global Search: Continuous Problems*. 1996 ISBN 0-7923-4238-0
14. P. Kouvelis and G. Yu: *Robust Discrete Optimization and Its Applications*. 1997 ISBN 0-7923-4291-7
15. H. Konno, P.T. Thach and H. Tuy: *Optimization on Low Rank Nonconvex Structures*. 1997 ISBN 0-7923-4308-5
16. M. Hajdu: *Network Scheduling Techniques for Construction Project Management*. 1997 ISBN 0-7923-4309-3
17. J. Mockus, W. Eddy, A. Mockus, L. Mockus and G. Reklaitis: *Bayesian Heuristic Approach to Discrete and Global Optimization*. Algorithms, Visualization, Software, and Applications. 1997 ISBN 0-7923-4327-1
18. I.M. Bomze, T. Csendes, R. Horst and P.M. Pardalos (eds.): *Developments in Global Optimization*. 1997 ISBN 0-7923-4351-4
19. T. Rapcsák: Smooth Nonlinear Optimization in R^n. 1997 ISBN 0-7923-4680-7
20. A. Migdalas, P.M. Pardalos and P. Värbrand (eds.): *Multilevel Optimization: Algorithms and Applications*. 1998 ISBN 0-7923-4693-9
21. E.S. Mistakidis and G.E. Stavroulakis: *Nonconvex Optimization in Mechanics*. Algorithms, Heuristics and Engineering Applications by the F.E.M. 1998 ISBN 0-7923-4812-5

Nonconvex Optimization and Its Applications

22. H. Tuy: *Convex Analysis and Global Optimization.* 1998 ISBN 0-7923-4818-4
23. D. Cieslik: *Steiner Minimal Trees.* 1998 ISBN 0-7923-4983-0
24. N.Z. Shor: *Nondifferentiable Optimization and Polynomial Problems.* 1998 ISBN 0-7923-4997-0
25. R. Reemtsen and J.-J. Rückmann (eds.): *Semi-Infinite Programming.* 1998 ISBN 0-7923-5054-5
26. B. Ricceri and S. Simons (eds.): *Minimax Theory and Applications.* 1998 ISBN 0-7923-5064-2
27. J.-P. Crouzeix, J.-E. Martinez-Legaz and M. Volle (eds.): *Generalized Convexitiy, Generalized Monotonicity: Recent Results.* 1998 ISBN 0-7923-5088-X
28. J. Outrata, M. Kočvara and J. Zowe: *Nonsmooth Approach to Optimization Problems with Equilibrium Constraints.* 1998 ISBN 0-7923-5170-3
29. D. Motreanu and P.D. Panagiotopoulos: *Minimax Theorems and Qualitative Properties of the Solutions of Hemivariational Inequalities.* 1999 ISBN 0-7923-5456-7
30. J.F. Bard: *Practical Bilevel Optimization.* Algorithms and Applications. 1999 ISBN 0-7923-5458-3
31. H.D. Sherali and W.P. Adams: *A Reformulation-Linearization Technique for Solving Discrete and Continuous Nonconvex Problems.* 1999 ISBN 0-7923-5487-7
32. F. Forgó, J. Szép and F. Szidarovszky: *Introduction to the Theory of Games.* Concepts, Methods, Applications. 1999 ISBN 0-7923-5775-2
33. C.A. Floudas and P.M. Pardalos (eds.): *Handbook of Test Problems in Local and Global Optimization.* 1999 ISBN 0-7923-5801-5
34. T. Stoilov and K. Stoilova: *Noniterative Coordination in Multilevel Systems.* 1999 ISBN 0-7923-5879-1
35. J. Haslinger, M. Miettinen and P.D. Panagiotopoulos: *Finite Element Method for Hemivariational Inequalities.* Theory, Methods and Applications. 1999 ISBN 0-7923-5951-8
36. V. Korotkich: *A Mathematical Structure of Emergent Computation.* 1999 ISBN 0-7923-6010-9
37. C.A. Floudas: *Deterministic Global Optimization: Theory, Methods and Applications.* 2000 ISBN 0-7923-6014-1
38. F. Giannessi (ed.): *Vector Variational Inequalities and Vector Equilibria.* Mathematical Theories. 1999 ISBN 0-7923-6026-5
39. D.Y. Gao: *Duality Principles in Nonconvex Systems.* Theory, Methods and Applications. 2000 ISBN 0-7923-6145-3
40. C.A. Floudas and P.M. Pardalos (eds.): *Optimization in Computational Chemistry and Molecular Biology.* Local and Global Approaches. 2000 ISBN 0-7923-6155-5
41. G. Isac: *Topological Methods in Complementarity Theory.* 2000 ISBN 0-7923-6274-8
42. P.M. Pardalos (ed.): *Approximation and Complexity in Numerical Optimization: Concrete and Discrete Problems.* 2000 ISBN 0-7923-6275-6
43. V. Demyanov and A. Rubinov (eds.): *Quasidifferentiability and Related Topics.* 2000 ISBN 0-7923-6284-5

Nonconvex Optimization and Its Applications

44. A. Rubinov: *Abstract Convexity and Global Optimization*. 2000 ISBN 0-7923-6323-X
45. R.G. Strongin and Y.D. Sergeyev: *Global Optimization with Non-Convex Constraints*. 2000 ISBN 0-7923-6490-2
46. X.-S. Zhang: *Neural Networks in Optimization*. 2000 ISBN 0-7923-6515-1
47. H. Jongen, P. Jonker and F. Twilt: *Nonlinear Optimization in Finite Dimensions*. Morse Theory, Chebyshev Approximation, Transversability, Flows, Parametric Aspects. 2000 ISBN 0-7923-6561-5
48. R. Horst, P.M. Pardalos and N.V. Thoai: *Introduction to Global Optimization*. 2nd Edition. 2000 ISBN 0-7923-6574-7
49. S.P. Uryasev (ed.): *Probabilistic Constrained Optimization*. Methodology and Applications. 2000 ISBN 0-7923-6644-1
50. D.Y. Gao, R.W. Ogden and G.E. Stavroulakis (eds.): *Nonsmooth/Nonconvex Mechanics*. Modeling, Analysis and Numerical Methods. 2001 ISBN 0-7923-6786-3
51. A. Atkinson, B. Bogacka and A. Zhigljavsky (eds.): *Optimum Design 2000*. 2001 ISBN 0-7923-6798-7
52. M. do Rosário Grossinho and S.A. Tersian: *An Introduction to Minimax Theorems and Their Applications to Differential Equations*. 2001 ISBN 0-7923-6832-0
53. A. Migdalas, P.M. Pardalos and P. Värbrand (eds.): *From Local to Global Optimization*. 2001 ISBN 0-7923-6883-5
54. N. Hadjisavvas and P.M. Pardalos (eds.): *Advances in Convex Analysis and Global Optimization*. Honoring the Memory of C. Caratheodory (1873-1950). 2001 ISBN 0-7923-6942-4
55. R.P. Gilbert, P.D. Panagiotopoulos† and P.M. Pardalos (eds.): *From Convexity to Nonconvexity*. 2001 ISBN 0-7923-7144-5
56. D.-Z. Du, P.M. Pardalos and W. Wu: *Mathematical Theory of Optimization*. 2001 ISBN 1-4020-0015-4
57. M.A. Goberna and M.A. López (eds.): *Semi-Infinite Programming. Recent Advances*. 2001 ISBN 1-4020-0032-4
58. F. Giannessi, A. Maugeri and P.M. Pardalos (eds.): *Equilibrium Problems: Nonsmooth Optimization and Variational Inequality Models*. 2001 ISBN 1-4020-0161-4
59. G. Dzemyda, V. Šaltenis and A. Žilinskas (eds.): *Stochastic and Global Optimization*. 2002 ISBN 1-4020-0484-2

KLUWER ACADEMIC PUBLISHERS – DORDRECHT / BOSTON / LONDON